Is the Human Species Special?

Why human-induced global warming could be in the interests of life

Neil Paul Cummins

Cranmore Publications

A catalogue record for this book is available from the British Library

Cover Image: Francesco Marino

ISBN: 978-1-907962-00-4

Published by Cranmore Publications

Reading, England

www.cranmorepublications.co.uk

What is man? – so I might begin; how does it happen that the world contains such a thing, which ferments like a chaos or moulders like a rotten tree, and never grows to ripeness? How can Nature tolerate this sour grape among her sweet clusters?

Friedrich Hölderlin

Contents

Preface

Amidst the hustle and bustle of daily existence there are sometimes moments when humans take a step back and reflect on some of the bigger questions relating to human existence. Do I have a purpose? How should I live my life? What, if anything, makes the human species unique? Does the human species have a purpose? Did the universe have a creator? In the light of the rapid spread of technology across the globe, what does the future hold for the planet Earth and its collection of wonderful life-forms?

These are questions which many of us spend at least a little of our life pondering. It is probably sensible not to spend too much time pondering these questions because there are no definitive answers to be found. However, it is probably fair to say that I have spent more time than most people pondering these questions and in this book I share with you my thoughts and the conclusions that I have reached.

If you read this book and are inspired, enthralled, informed or entertained, then that is fantastic. I am writing this book in the hope that this will be so. There is also a more serious side to the book. I put forward a particular view of the place of the human species within an evolving universe, and this view gives rise to some suggestions as to how the Earth might be made a better place to live. Firstly, I make a suggestion as to how the human species can minimise the negative effects of global warming. Secondly, I make a suggestion as to how individual humans could think of and treat the other life-forms on the Earth. Thirdly, I make a suggestion as to how you might be able to attain a more satisfying and healthy life.

The view of the human species that I put forward is concordant with a very broad range of religious, spiritual and scientific beliefs. You could think of the view that I put forward as an overarching view within which these various beliefs are situated. Furthermore, it is possible that the view that I put forward could give a new perspective to your existing beliefs. So, I would hope that as you read the following chapters you bear this in mind and take a moment to consider how what you are reading could be consistent with your existing beliefs.

Introduction

Many books have been written on human-induced global warming; it is one of the most important issues in human society in the early twenty-first century. Many books have also been written on whether the human species is special; this is an issue which humans have been pondering for many millennia. My aim in this book is to bring these two issues together. I will be suggesting that these two issues are fundamentally interconnected. I will be proposing that the human species is special and that this specialness is deeply connected to human-induced global warming.

When thinking about these two issues it is very helpful to keep in mind that the universe is an evolving entity. For the sake of clarity I should make it clear that I am using the word 'universe' to refer to everything that humans can possibly perceive. So, humans *actually* perceive the Earth, the other planets in our solar system, other solar systems, our galaxy and a plethora of other galaxies. This means that all of these entities *and* all of the individual parts of these planets, solar systems and galaxies are part of the 'universe'. All of the entities that exist which humans have not yet perceived, *but could possibly perceive,* are also part of the 'universe'.

Why do I not simply use the word 'universe' to refer to everything that exists? I don't do this because I wish to leave open the possibility that everything that humans can possibly perceive is not the same as everything that exists. By this I mean that the 'universe' (everything that humans can possibly perceive) could be an enclosed 'bubble' which is surrounded by other entities. If this is so, then it is possible that the universe was created by an entity that was located outside of the universe; this entity could have created the enclosed 'bubble'.

So, the universe is an evolving entity. Just like you have a birth, a period of ageing, and a coming to an end, so do planets, solar systems and galaxies. The constituents of the universe are always evolving. Of particular interest to us is that part of the evolving universe which is the planet Earth. The Earth had a birth and is now ageing. As all parts of the universe are evolving the Sun is also ageing; the importance of this will become clear. The ageing of the Sun and the ageing of the Earth are tightly connected. It is helpful to keep in mind that it is at a particular point in this tightly connected ageing process that the human species came into existence.

In the first four chapters our focus is on the question of whether the human species is special. If the human species is *not* special then it would be just another species of animal. In *Chapter One* I reflect on the fact that most humans have a sense that the human species is special and I note the important difference that exists between having a sense that one's species is special and one's species actually being special.

I contend that there is a singular source of the sense of human specialness; however, I suggest that most humans believe that the human species is special because they believe that humans have at least one of a range of unique attributes. This issue of human uniqueness is the focus of *Chapter Two*. What, if anything, is unique about the human species? Is there an attribute that humans have that the rest of the planetary life-forms lack? And, even if such an attribute exists, is this of any importance? I propose that the human species has unique attributes but that there are non-human species which have unique attributes too. Furthermore, I conclude that the human species does not have a 'special' unique attribute which *in itself* can justify the sense that the human species is special.

Having proposed that the possession of a unique attribute does not make the human species special I then turn to the question of whether the human species has a purpose; this is the focus of *Chapter Three*. In the context of the evolution of planetary life over billions of years does the human species have a purpose to fulfil that only it can fulfil? If such a purpose exists then how does this relate to the question of whether the universe had a creator? Could everything that exists have a purpose? Or, could it be the case that only some things have a purpose? I propose that the human species has a purpose, but that there are also purposes throughout the universe. This means that the human species having a purpose does not *in itself* make the human species special.

So, is the human species special? I consider this question in *Chapter Four*. What does it mean to be special? Let us assume that numerous species of animals have unique attributes *and* that everything has a purpose; does it then follow that the human species is just another species of animal? No. In this chapter I propose that the human species is the pinnacle of planetary evolution; it is the species which planetary life has been striving to attain since it first arose. The human species has the most important purpose there could possibly be – to be the saviour of planetary life.

In *Chapter Five* our focus switches to the environmental crisis. If the human species is the saviour of planetary life why is it doing so much damage to planetary life? One will probably be wondering how deforestation, the plundering of the seas, acid rain, oil leaks in the oceans and human-induced global warming could possibly be in the interests of planetary life. In this chapter I outline the important differences between the 'environmental crisis', 'climate change' and 'global warming'. These three phenomena are closely related but importantly distinct. I suggest that it is *human-induced global warming* which is in the interests of planetary life. This is not to say that in itself global warming is a good thing; this is far from the case. Rather, it is human-induced global warming that is playing an important role in the fulfilment of the purpose of the human species.

Having suggested that the human species is special because it has a vital purpose, and having considered how this relates to the environmental crisis, I then turn to the question of how individual humans come to fulfil the purpose of the human species.

In *Chapter Six* I consider whether individual humans have a purpose. How is it possible that individual humans can act in such a way that the purpose of the species gets fulfilled when these individuals don't realise that this purpose exists? I give an account in which the vast majority of humans alive at any moment in time each play a part in the fulfilment of the purpose of the human species without realising that they are doing so.

In *Chapter Seven* I consider why humans perceive parts of the universe to be beautiful. Could it have transpired that humans perceive the universe to be wholly devoid of beauty? How does seeing the universe as beautiful relate to individual humans acting in a way which results in the fulfilment of the purpose of the human species? Addressing these questions is the focus of this chapter.

Given the view that I have put forward, how should one think of and treat the non-human life forms which we share the planet with; this is the focus of *Chapter Eight*. I contend that it was inevitable that humans would come to see these life-forms simply as resources for human use. However, I also contend that there is absolutely no need for a human alive today to see non-human life-forms as resources for human use. Rather, I urge one to see them as companions, and to treat them in a respectful manner.

Having covered a lot of ground, I then discuss my conclusions which draw together all of the strands of thought from the previous chapters; these conclusions are to be found in *Chapter Nine*. Finally, in the *Introduction to the appendices* I explain why I have included two appendices in the book.

Chapter 1

A sense of specialness

It is surely true that the overwhelming majority of humans alive today have a sense that the human species is special. By this I mean that the overwhelming majority of humans alive today have a sense that the human species is *not* just another species of animal; they have a sense that there is a fundamental division of some kind with the human species on one side and all of the other planetary life-forms on the other side. Where does this sense of specialness come from? If one goes far enough back in the life of a human then one will arrive at the moment when this sense of specialness first arose in that human. It seems reasonable to assume that human infants do not have this sense of human specialness. When human infants explore their surroundings they don't have a sense that the human species is somehow special, somehow divided from the other planetary life-forms.

Perhaps it is when a human reaches what one can loosely call a 'broad overview' of their surroundings that this sense of specialness arises. A 'broad overview' of one's surroundings entails a wide range of knowledge about those surroundings; not detailed knowledge about any particular thing, just a little knowledge about lots of things. So, if one knows that there are countries and continents, mountains and oceans, fish and birds, insects and bacteria, planets and stars, boats and cars, aeroplanes and space stations, apes and dolphins, mobile phones and laptop computers, and cats and dogs, then one has a 'broad overview' of one's surroundings.

It seems almost inevitable that when a human first has a 'broad overview' of their surroundings that they will have a sense that the human species is special. This is because humans seem to have abilities that other things lack. If a human grows up surrounded with designer clothes, a nice house, books, cars, aeroplanes, skyscrapers, mobile phones and computers, then it is not surprising that a sense of human specialness arises within that human. After all, no other species of animals can be found reading books, flying aeroplanes, or texting on their mobile phones.

It is very easy to understand why a human can have *a sense* that their species is special. But is the human species *actually* special? And, if it is, *why* is it special? For, when one thinks about it, it is also very easy to understand how members of a non-human species could have *a sense* of specialness too. Let us consider an eagle. As the eagle flies majestically through the sky it looks down and sees humans who are rushing around in their 9 to 5 working day, stuck in traffic jams, and jostling their way through crowds. It is easy to imagine how the eagle, who spends its day gliding around in the sky, could have a sense that it, and other eagles, must be special. Let us consider a dolphin. As it communicates through echolocation, as it jumps out of the ocean and spins around in the air, as it saves a human from near-certain death, it could, surely, have a sense that it, and other dolphins, must be special.

What does this mean? It means that a sense of specialness and actually being special are two different things. It could be the case that the human species is special. However, if the human species is special this cannot be just because individual humans have a sense that their species is special.

Chapter 2

Does the human species have a unique attribute?

I have claimed that the overwhelming majority of humans alive today have a sense that the human species is special – a sense that there is a fundamental division of some kind between the human species and all of the other planetary life-forms. I am convinced that the sole source of this sense of human specialness is the advanced tool-using ability of the human species. It is because humans are born into, and grow up in, surroundings which are pervaded with *advanced human tools* that the sense of human specialness arises. If I am right, then this means that in the early stages of the evolution of the human species there would not have been a widespread sense of human specialness. I would suggest that the tools used by hunter-gatherers – bow and arrow, harpoon, atlatl, projectile points – are advanced enough to give rise to a sense of human specialness.

What, exactly, is an advanced tool? I take the following to all be examples of advanced tools: knives, forks, buckets, tee-pees, cups, aeroplanes, harpoons, books, submarines, laptop computers, cars, televisions, pens, bicycles, skyscrapers, spacecraft, tables and chairs; some of these advanced tools are clearly more advanced than others. These tools are 'advanced' compared to 'non-advanced' tools such as twigs and stones. 'Non-advanced' tools are still used by humans today and they were the only tools used by early humans. I am proposing that when the human species started to modify its surroundings in more than the simplest of ways that this resulted in a wide-spread sense of specialness within the human species. *The source of the widespread sense of human specialness which exists today is advanced tool use.*

You might not be convinced by this – you might believe that your sense that the human species is special has an alternative source. For instance, you might be thinking: *if all of the advanced tools on the Earth disappeared I would still have a sense that the human species is special.* And, of course, you would be correct. You have grown up in surroundings which are pervaded with advanced human tools and therefore have a sense of human specialness; once this sense of human specialness has arisen it will not vanish simply because all of the advanced tools on the Earth have disappeared.

What we are interested in is whether a human who is born into, and grows up in, surroundings which are wholly devoid of advanced human tools would have a sense that the human species is special – a sense that there is a fundamental division between the human species and all of the other planetary life-forms. On the Earth today it would be hard to find such a human. There are still small groups of humans who live on secluded islands and deep within rainforests but these groups use advanced human tools. There is always a chance that a true 'feral' child will be found – a human who from birth was raised by non-human animals in the wild. I am fairly convinced that such a human would not have a sense that the human species is special.

It seems that there aren't any humans alive today who haven't been exposed to advanced human tools so our only option is to engage in a thought experiment. In order to rid yourself of the effects of advanced human tools you need to imagine not only that the Earth is devoid of these tools, you also need to imagine that you are a human who is living in the distant past. You need to imagine that you are living at a time before humans developed the advanced tools of the hunter-gatherers. Can you imagine what it would be like to be a human living at such a time? When I try to imagine what it was like for these early humans the last thing that I imagine is that these humans had a sense that they were special – that they were fundamentally divided from the other planetary life-forms. Rather, I imagine that these humans were regularly killed by non-human animals; that there was a precarious battle for survival; that they lived in a state of fear, of vulnerability. Compared to the killing ability of many of the species of animals which existed in this period the human species was physically inferior – humans were relatively small, soft-skinned and small-toothed; numerous other species had a physical constitution much more suitable for killing. Without advanced tools to defend themselves it seems possible that these humans might have had *a sense that their predators were the life-forms which were special.* However, it is perhaps more likely that these early humans had neither a sense that their species was special nor a sense that any other species was special. If there are no advanced human tools then a sense of human specialness doesn't arise. So, I'll repeat what I said a little earlier. *The source of the widespread sense of human specialness which exists today is advanced tool use.*

There is an important difference between the *source* of the widespread sense of human specialness and the *explanations* that individual humans come up with to justify their sense that the human species is special. Humans are rational animals; in addition to having intuitive 'senses' they also try to rationalise things. It is interesting that when humans rationalise their sense of human specialness they typically do not conclude that humans are special because they are advanced tool-users. Some humans do reach this conclusion. However, if one asks a large number of humans why they believe the human species to be special one will typically get answers such as: humans are the only life-forms that have souls;

humans are the only life-forms that have language; humans are the only life-forms that can think; humans are the only life-forms that can feel pain/have emotions; humans are the only moral life-forms; humans are the only life-forms that have awareness; humans are the only life-forms with culture. So, it seems that most humans believe that the human species is special because they believe that humans have at least one of a range of unique attributes.

So, whilst the *sense of* human specialness arises from advanced tool use, most humans *rationalise* that the human species is special for a completely different reason. In this chapter my aim is to open you to the possibility that none of the rationalisations (including advanced tool use) that humans typically make about why the human species is special are true. In other words, I am contending that the widespread belief that the human species is special because it has a unique attribute is false.

It is true that the human species has unique attributes, but having unique attributes does not make a species special. This is because there are a plethora of species which have unique attributes. For example, whales are uniquely able to communicate half way around the world via their songs. Hummingbirds are uniquely able to hover in the air by being able to flap their wings 80 times per second. And sharks are uniquely able to detect electrical pulses in the ocean up to a mile away because of their special brain cells.

Our concern is whether the human species is special – whether there is a fundamental division between the human species and the rest of the planetary life-forms. For a unique human attribute to give rise to such a division it would have to be a really special attribute. Such an attribute would need to warrant talk of a *great chasm with the human species on the one side and the rest of the planetary life-forms on the other.* If humans are the unique possessors of such a 'special' attribute then the human species would be special; if there is no such attribute then the widespread rationalisation that such an attribute makes the human species special is wrong. In order for there to be a chasm it needs to be the case that the human species is the only planetary life-form to have the attribute. If humans have an attribute to an advanced degree and a non-human species has the same attribute to a less advanced degree then there is no chasm between the two species; there are simply different degrees of development of the same attribute.

Let me say a few words about how the human species could have acquired such a 'special' unique attribute. If one believes that the universe had a creator then one could believe that this creator uniquely endowed the human species with a 'special' attribute. Whilst, if one does not believe that the universe had a creator, then one could believe that it is solely the unique evolutionary history of the human species which has resulted in humans being the only planetary life-form to have evolved this 'special' attribute. I should make it clear that I don't have a particular view as to whether or not the universe had a creator (how could I possibly know?) but I do believe, and will assume, that the universe is an evolving entity and that the human species evolved from other planetary life-forms.

Now, let us consider all of the obvious possible candidates for such a 'special' unique human attribute. One of the most obvious candidates is the soul. One could believe that the human species is special because all humans have a soul *and* no non-human planetary life-forms have a soul. If this were true then clearly the human species would have a very 'special' unique attribute; the soul would be an attribute which causes there to be a great chasm; on one side of the rift is the ensouled human species, on the other side of the rift is all of the other planetary life-forms. So, is this true? Does such a chasm exist?

This is a hard question to answer. Let me start with myself. I am a human. If someone tells me that I have a soul I don't know whether I should believe them or not. From my everyday experience I don't have a clue as to whether or not I have a soul. The point is that, as far as I know, no human has ever seen a soul, and nobody truly knows whether such a thing exists. This is, perhaps, why the exact nature of what the soul is meant to be is unclear.

Let us accept the possibility that every human has a soul. I take this possibility to entail that there are two parts to every human – a biological part and another part which still exists when the biological part ceases to function. (If you have a more specific idea about what the soul is that is fine – the following considerations will still apply.) If every human has a soul does it follow that there is a great chasm – the soul-possessing human species versus the rest of the planetary life-forms? Of course it doesn't. If one accepts that the human species evolved from a different species, then one has a serious issue to consider. If humans have souls then does the species which humans evolved from also have souls? If one accepts that the human species evolved out of a different species, and also believes that every human has a soul, but denies that the precursor species had souls, then one has some explaining to do. One needs to assert that at one moment all of the species on the planet existed without any souls, and the next moment, when the first human evolved, that they were somehow accompanied by a soul. Of course, one could have faith that this just happened, and if one has such faith one could be correct. However, if one believes that all humans have souls, and also takes the evolution of the human species from another species seriously, then from a rational perspective it is much easier to believe that the precursor species also had souls. These considerations will lead some people to conclude that maybe humans do not have souls; these people would rather deny that humans have souls than accept that non-humans have souls.

In conclusion, whatever exactly the soul is I am happy to accept the possibility that I have one on the condition that at least one non-human species also has souls. This seems to be the most reasonable position one can take on the matter. If one agrees with me then one will also conclude that the possession of a soul is not a 'special' unique human attribute. Either humans are not the only species to have souls, or there is no such thing as a soul.

Which other attributes could one believe to be uniquely human and 'special'? In the recent past many humans believed that the ability to use tools was a unique human attribute. However, there is now a plethora of evidence that numerous non-human species use tools. Some of the non-human animals which have been observed using tools are primates, birds, elephants, dolphins, otters and octopuses. It is obviously true that many non-human species use tools; therefore, tool-use is not a 'special' unique human attribute. Of course, it is also obviously true that the human species can use tools in a much more advanced way than any other species. However, when there are various degrees of advancement of an attribute in different species this means that there is no great chasm between those species.

What about the ability to feel pain or to have emotions? Again, as with tool-use, in the past these were attributes that many humans seemed to believe were unique attributes of their species. Anyone who has spent any length of time observing cats, dogs, chimpanzees, and numerous other species, will surely find this position hard to accept: *it is blindingly obvious that these creatures feel pain and have emotions* these people will exclaim. This view – that pain and emotions are not uniquely human – is given rational backing if one accepts that the human species evolved out of another species; I find it very hard to accept the idea that the first human could feel pain and have emotions but that its non-human ancestor was wholly painless and emotionless. Of course, one could insist that the species that gave rise to the human species was wholly painless and emotionless. However, common sense does not accord well with this insistence. Let us accept therefore that the ability to feel pain and the ability to have emotions are not 'special' unique human attributes.

What about the ability to think, to rationalise? It is surely the case that dolphins, chimpanzees, bonobos, and a plethora of other species think. These non-human animals have been observed solving novel problems and acting in ways that can only reasonably be explained if they are thinking. Indeed, one reasonable conclusion to reach would be that every animal that has a brain thinks because a brain is a thinking thing. Of course, our present concern is only to note the likelihood, bordering on certainty, that there is one animal that is not human that thinks. Given that this is so, it means that thinking is not a 'special' unique human attribute.

Let us briefly consider the relationship between advanced thought and a species being special. After analysing the brain structure of dolphins some people have claimed that dolphins have a more advanced ability to think than humans. It seems that both dolphins and humans have a very advanced ability to think. I suggest that the question of whether the human species is special is not dependent on whether humans have a more advanced ability to think than dolphins. I can quite happily accept the possibility that dolphins have a more advanced ability to think than humans. However, this acceptance will not cause me (and, I hope, it would not cause you) to believe that the human species is not special. Having the most advanced ability to think on the Earth does not make a species special.

Having considered the attribute of thought let us now consider the attribute of awareness. Awareness is a slippery phenomenon. This is because whilst one can be sure that oneself is aware it is often very difficult to judge whether or not other life-forms have the attribute. One will probably have heard of cases where people in hospital have *been aware of* their doctor and relatives discussing whether to turn off their life-support machine because there was no chance that they would *regain* awareness! Then there is the phenomenon of 'blindsight' which affects people who have a damaged primary visual cortex; these people are unaware of objects which are in front of them, yet when forced to guess they are able to give information, at levels significantly above chance, about the objects. This implies that there can be perception without awareness. Another example of perception without awareness is somnambulism (sleepwalking). A sleepwalker can *appear* to be aware of their surroundings because they are successfully navigating those surroundings and performing apparently thoughtful activities; they may even utter the words "I am aware". However, *in reality,* sleepwalkers typically claim that they were totally *unaware* whilst sleepwalking.

It might seem obvious to you that non-human animals such as cats, dogs, dolphins, chimpanzees, and a plethora of other species, have the attribute of awareness. And if it does you are probably correct. However, given the 'hospital', 'blindsight' and 'somnambulism' cases the possibility that non-human animals could be thinking and perceiving without awareness should be briefly considered.

Even if it were the case that some non-human animals can only think without awareness it only requires a single non-human life-form to have the attribute of awareness to mean that awareness is not a 'special' unique human attribute. When we consider the evolutionary perspective, the evolution of humans from non-human animals, this might cause one to conclude that it is likely that our non-human ancestors had the attribute of awareness. However, it is at least a possibility that the first humans themselves did not have the attribute of awareness; they could have been thinking without awareness; awareness could have arisen in humans a long time after the evolution of the species. Nevertheless, even if this were so, it is still likely that other species of animals have also evolved the attribute of awareness. So, the common ancestor of both humans and chimpanzees could have not had the attribute of awareness, and the first humans and chimpanzees could have not had the attribute of awareness, but both species could have gained the attribute at a later time.

Let us conclude. Taking everything into account I suggest that it is highly unlikely that the human species is the only species of animal that has the attribute of awareness. Furthermore, *even if* awareness was currently a unique human attribute, one has to accept the possibility that in the future a non-human planetary life-form could have the attribute of awareness. If this came to pass, and awareness was no longer a unique human attribute, would one then assert *oh well, the human species is no longer special.* I hope one would not assert such a thing. Rather, I hope one agrees with me that the human species is special irrespective of whether or not a non-human planetary life-form also has the attribute of awareness. The attribute of awareness *does not* make the human species special. (If, unlike me, you believe that self-awareness is a *different* attribute to awareness, then what has been said here with regards to awareness also applies to self-awareness.)

Which other attribute could one believe to be a 'special' unique human attribute? The ability to use language is another possibility. In order to justify their sense of human specialness it is common to hear people assert that humans have language but that all non-human species only have the ability to communicate. This assertion is, no doubt, heavily influenced by the fact that humans have nimble fingers to write with, printing presses to produce books and newspapers with, and a plethora of libraries. If divers were to find a dolphin library then one would probably conclude that there was a dolphin language! Of course, dolphins don't have fingers with which to write or underwater printing presses, so such a library won't be found. But, the lack of such things obviously doesn't mean that dolphins don't have a language.

It is undeniable that a multitude of species of animals communicate in very complex ways. The fact is that we are largely incapable of understanding their communications systems. If humans were capable of learning the communication systems of non-human animals, so that we could fully communicate with these animals in their 'language', then it would perhaps be obvious that some of these species do have their own language. A couple of years ago I went to a foreign country where I didn't speak the language and I felt uncomfortable when I was surrounded by humans who were making lots of strange sounds that I couldn't make any sense of. I guess it might have been a language; but it could have been that I was surrounded by people who were making sounds that were utterly meaningless.

When one thinks about it, it is pretty amazing that there are some non-human animals that are able to communicate with humans in the *human* language. Two of the most famous such non-human animals are Alex and Koko. Alex was an African Grey parrot who was able to speak and manipulate the English language. Koko is a gorilla who lives in California; he has mastered 1000 words in American Sign Language and is able to combine them in novel ways. Furthermore, it should be kept in mind that chimpanzees have been observed using symbolism within *their* societies.

Humans do seem to have a very advanced language, and to have the unique ability to store this language in a diverse range of ways so that it can be accessed by future generations (these ways include things such as books, blogs, and DVDs). However, I assume that one will agree with me that there is no great chasm with the language-using human species on the one side, and the rest of the planetary life-forms on the other. Humans, perhaps, simply have a more advanced ability to use language than some other species. In other words, perhaps the human language contains a lot more words/expressions than the languages used by some non-human species.

How about morality? In the recent past the view that the human species is the only moral species has been dominant. However, there is now ample observational evidence that some species of non-human animals have the attribute of morality – they have a sense of fairness, a sense of right and wrong. Primatologists, such as Professor Frans de Waal, have concluded that acts of consolation and empathetic behaviour are universal amongst the Great Apes. The existence of morality in primate societies should not surprise one. After all, group living entails shared values and requires individuals to take into account the needs of other members within the group. In other words, individuals within a society need to have a sense of what is right and what is wrong. It is likely that morality exists in a number of species of non-human animals. However, the existence of the attribute in a single non-human species means that it is not a 'special' unique human attribute.

Finally, let us consider culture. What exactly is culture and is it something that is uniquely human? Culture is a term that humans typically use to refer to some of their activities. Indeed, it is a term that has traditionally been used to create a separation between humans and non-human animals: *culture is what humans have and non-humans don't*. This obviously won't do. The actual activities of humans that are supposedly 'cultural' activities need to be specified so that other species can be closely monitored to see if they also partake in these activities. Such observations have revealed that symbolism, tool use, social conformity and learned behaviour exist in chimpanzee societies. So much evidence has been amassed that now numerous primatologists and many anthropologists have concluded that chimpanzees have culture too. As the monitoring of other species continues in the future it is entirely possible that several other species will also be identified as having culture.

One should not be concerned by the evidence that chimpanzees have culture. All this means is that culture cannot be a 'special' unique human attribute which generates a great chasm between the 'cultured' human species and the rest of the planetary life-forms. Culture, and human culture specifically, is such a wonderful thing. Human culture does not get denigrated in any way by the acceptance that chimpanzees have culture. Human culture is so much more advanced than chimpanzee culture. Nevertheless, the attribute of culture, *in itself,* cannot be said to be a 'special' unique human attribute.

So, with the list of possible 'special' unique attributes which could make the human species special exhausted, our conclusion is that no such special attribute exists. In other words, there is no great chasm with the human species on the one side and the rest of the planetary life-forms on the other side. The human species does have unique attributes; as we have seen two such attributes are an advanced ability to use tools and an advanced culture.

However, these attributes do not make the human species special because non-human planetary life-forms have these same attributes in a less advanced form.

It is possible that you might disagree with this conclusion. You might believe that there is an attribute which makes the human species special that I have omitted, or that one of the attributes I have considered is uniquely human. Let us assume that one of the attributes I have considered *is* actually uniquely human. There is a fundamental issue which we haven't addressed yet. I have considered the attributes which humans typically rationalise as being the attributes that make the human species special. But, why should these attributes *actually* be special? Why should these attributes be 'special' ones whilst the unique attributes of non-human animals are not special?

I suggest that unless life has a purpose there are no special attributes. There are simply *different* attributes. Every species has a bundle of different attributes; some of the attributes in this bundle will be advanced *compared to the degree of advancement in other species* and some will not. Unless life has a purpose (in this chapter we have been considering the attributes on their own merit and assuming that life does not have a purpose) there is no good reason for saying that a unique human attribute is a 'special' attribute; so, there is no good reason for saying that the human species is special.

Of course, it is possible that the human species could be special *not* because humans possess a 'special' unique attribute, *but* because the human species has a special purpose. If this were the case then there would be a sense in which the unique human attributes which enable the fulfilment of that purpose *are* themselves special. Before we can consider whether the human species has a *special* purpose we need to consider whether the human species has a purpose.

Chapter 3

Does the human species have a purpose?

What would it mean for the human species to have a purpose? Let us not confuse this question with the question of whether the universe had a creator. Clearly, if the universe had a creator then the creator could have endowed the human species with a purpose; that is to say, the creator could have designed the universe so that when the human species evolved it had a purpose to fulfil. But even if the universe did not have a creator the human species could still have a purpose. To have a purpose is simply to have a goal, a mission, a target, an objective. If the human species has a goal then the human species has a purpose.

It is possible that *everything* in the universe has a purpose. It is also possible that *nothing* in the universe has a purpose. A third possibility is that some things have a purpose and some things do not. If purposes are widespread in the universe then the questions we are really interested in are: Does the human species have a special purpose? Is the human species special?

Let us consider the possibility that the human species has a purpose. It is possible that you might find this idea to be slightly fanciful. *How could an entire species have a purpose?* Where does the purpose come from? Who could possibly be ensuring that this purpose gets fulfilled? Who is doing the organising? These are understandable concerns, but a species can fulfil a purpose without any member of that species realising that they are doing so. It is quite easy to think of situations in which humans use other humans in a way that these 'used' humans fulfil a purpose without them realising it. The same principle can apply to a whole species. The individual actions of all of the members of the human species over a certain period of time could result in the human species fulfilling a purpose without any of the individuals realising that the purpose existed in the first place. The way in which this process of unaware purpose-fulfilment could work will be considered in later chapters, as will the issue of where the purpose comes from.

This possibility of unaware purpose-fulfilment in the human species has wider implications. The human species as a whole can fulfil a purpose without any human knowing about the purpose or thinking about the purpose. If a purpose can be fulfilled in this way then it is easy to envision how it is possible that entities such as trees could have a purpose, how cats could have a purpose, and how fish could have a purpose. In short, one can envision how everything could have a purpose.

Let us consider the possibility that everything has a purpose. The existence of purpose within everything is perhaps easier to envision if one believes that there was a creator who created everything and gave everything a purpose. However, even if there was no creator, one can still make sense of the idea that everything has a purpose. Let us assume that there was no creator and consider the various parts of the human body. The purpose of the heart is to pump blood around the body, the purpose of the stomach is to enable the digestion of food, and the purpose of the lungs is to enable a human to breath. Now, if one imagines that the entire planet is one's body, then one can envision the purpose of the individual parts of the Earth. The snow in the Arctic and the Antarctic has the purpose of raising the planetary albedo in order to prevent the planetary temperature from rising too much (the planetary albedo is the amount of sunlight reflected, rather than absorbed, by the surface of a planet; snow has a high reflectance level which results in a lower temperature); plants have the purpose of turning sunlight into energy sources which can sustain animals; the water cycle is the lifeblood of the planet whose purpose is to sustain all living things. In this way one can envision how everything on the Earth has a purpose. One can then imagine that the universe is one's body, the individual parts of which interact in such a way so as to fulfil the purpose of creating planets on which life can arise.

What exactly does it mean to assert that everything in the universe has a purpose? What it means is that the entire universe is *pervaded* with purpose. In other words, the smallest constituents of the universe (which we can call 'atoms') have a purpose and arrangements of atoms have a purpose too. To see this let us consider a particular human. The atoms of their body have a purpose, their cells have a purpose, their organs have a purpose and they as an individual human have a purpose. For this reason the view that everything has a purpose is best thought of as a 'multi-level' view of the universe. Purposes pervade the universe at many different levels – atoms, organs, individual life-forms, species and life as a whole.

This is a vision of the universe as an entity which *strives to fulfil purposes* over varying periods of time. Some purposes will be strived for and fulfilled over a very short-period of time. Whilst other purposes will be gradually strived for over very long periods of time – tens of thousands of years, hundreds of thousands of years, millions of years. It is undoubtedly hard to imagine how an entity can gradually strive for something over a million years when one's own existence is one of 24 hour days, 7 day weeks, and 100 years in total if one is very lucky. We live for such a short time that it is hard to grasp such a thing. One can easily understand how one can strive for something over a 40 year period. One can save a set amount of money each and every day, and after 40 years one will have fulfilled the purpose of paying off one's mortgage. The increase in savings each day will be minimal, but after 40 years the cumulative effect is easy to see. If one can try to imagine the Earth striving to fulfil a purpose not over 40 years, but over 1,000,000 years, then one will surely appreciate that whilst the end products of universal strivings – purpose fulfilment – may be easy to perceive, the whole process of striving itself can be imperceptible to a human.

So, on this view, according to which everything is pervaded with purpose, one *can* perceive (with one's senses, and typically with one's eyes) some of the strivings and some of the purposes being fulfilled. For example, one can perceive parts of the water cycle such as the movement of clouds, precipitation and flowing streams. If the entire universe is pervaded with purpose, then *everything* one perceives contains strivings and purposes which are being fulfilled (atoms are striving and fulfilling purposes as are a multitude of their arrangements such as bodily organs). So, when one perceives a human who is performing a particular activity this human will contain a plethora of strivings and purposes; the activity they are performing will also arise from one of their individual strivings to fulfil a particular individual purpose. Furthermore, it is possible that the activity they are performing is also partially fulfilling the purpose of the human species; however, this need not be the case. We will return to the issue of individual human purpose in *Chapter Six*.

However, whilst one can perceive some strivings and purposes being fulfilled, the situation is more complex when it comes to very long processes of striving. Let us imagine that the human species has a purpose and that planetary life was striving for millions of years to bring the human species into existence. Would it be possible to perceive this process of striving?

Before answering this question I should make it clear what I mean by 'planetary life'. The phrase does *not* refer to *individual* life-forms; rather, it refers to the *totality* of life-forms that exist on the Earth at a particular moment in time. So, if when one reads the phrase 'planetary life' one thinks of a particular cat, or a species of cats, then one has missed the meaning that the phrase is meant to convey. One should not even think of 'planetary life' as a *collection of individual* life-forms; one should think of it as *a single entity* that is constituted out of individual life-forms. It is perhaps helpful to imagine that the Earth is divided into two parts – you can imagine that all planetary life-forms are coloured green, whilst all non-living parts of the planet are coloured red; the green part will be 'planetary life'. I will also sometimes use the phrase 'life' to refer to the *totality* of life-forms in the universe; 'life' will be equivalent to 'planetary life' if the only life-forms in the universe exist on the Earth.

Let us return to our 'striving for millions of years' scenario. In this scenario the complete process of striving is exceptionally difficult to perceive. A perceiver would need to perceive planetary life for millions of years in order to be able to appreciate the whole process of striving which culminated in the purpose being fulfilled. So, if an entity which had the lifespan of a human was able to continuously perceive this process for their entire life they would only perceive a tiny fraction of the process (100 years of a process that takes millions of years). From perceiving this miniscule fragment the perceiving entity wouldn't know that what they had perceived was part of a process of striving which had a timescale of millions of years. This means that the universe could be pervaded with processes of striving which one is unable to meaning-fully perceive.

We have been considering the possibility that every-thing has a purpose. Let us consider the possibility that some things have a purpose and some things do not. Where would one draw the division? Perhaps one might think that the living has a purpose and the non-living does not. This is a plausible thing to believe. However, it is worth remembering that the boundary between the living and the non-living is a fuzzy boundary; there are entities for which it is not clear whether they are 'living' or 'non-living'. Of course, as long as there is such a thing as a distinction between the living and the non-living it is not necessary to know where the exact boundary is; one can just assert that the living has a purpose and the non-living does not. There is some intuitive appeal in the idea that living things have a purpose whilst non-living things do not. Nevertheless, if one accepts that all living things have a purpose then there is something to be said for believing that *everything* has a purpose; this view of the universe is more holistic, more complete.

There is another way in which one could envision the possibility that some things have a purpose and some things do not; one could have a *multi-level view* of the universe and hold that only certain levels have a purpose. One variant of this view is that 'higher levels' – such as the human species as a whole – have a purpose, but 'lower levels' – such as the bodily organs of a human – do not. However, the multi-level view is most likely to be used in the opposite way, so let us focus on this opposite possibility. On a multi-level view the organs that comprise a human body could have a purpose – the purpose of the heart is to pump blood – but the individual whose body it is could have no purpose and the human species as a whole could have no purpose. From our perspective this view is closely related to the view that nothing has a purpose; on the 'nothing' view the human species has no purpose and life has no purpose; there is no purpose anywhere. Whilst on this multi-level view it is also the case that the human species has no purpose and life has no purpose; the difference with this view is that entities such as a heart do have a purpose. Both of these views reject the idea that the human species has a purpose.

So, does everything have a purpose, nothing have a purpose, or do only some things have a purpose? There is no conclusive answer to this question. One has to consider the world around one, consider the points that I have made in this chapter, and take a view – everything/ nothing/some things/don't know. One is largely forming a view from a position of ignorance. As we have seen, the human lifespan is far too short to be able to perceive long processes of striving. And, more fundamentally, even if one could perceive such processes one would not know for certain what one was observing. This is because the *motivation* to strive to fulfil a purpose is something that comes from *within* an entity. When one observes the actions of another human over a certain length of time one can often deduce that within that human there is an objective which that human is attempting to fulfil; in other words, one can infer that the human is striving to fulfil a purpose. But, importantly, one *cannot* observe the cause of this striving – the motivating states which are causing this human to strive. One can possibly deduce what the motivating states might be, but one cannot observe these states; the motivation to strive is something that is internal to the human and it precedes their observable actions. One simply observes another human and one infers from their actions that there must be some internal states within that human which are causing them to strive to fulfil a purpose.

Why is this important? Clearly, if one cannot observe the internal states within a human that are motivating that human to strive to fulfil a purpose, then one is also unable to observe the states within a tree (if such states exist) that are motivating the tree to strive to fulfil a purpose. One's perceptions of the world are limited – one perceives external movements of things and is barred from their interior. A more familiar example of this is the pain of another human. When one sees a human with blood running down their arm who is screaming one will, no doubt, unless one believes the human to be an actor, infer that there are states of pain *within* this human. However, one obviously cannot perceive the states of pain them-selves! These are *internal* to the human one is observing. One can perceive some of the external manifestations of the internal state but one is perceptually barred from the internal state itself. So, in short, one does not know for certain whether there are internal states such as pain within other humans, and analogously, one does not know for certain whether there are motivations to strive within other humans. One probably presumes that such states do exist because of one's observations and because of one's own experiences of these states and motivations to strive within oneself.

So, are there internal motivations to strive to fulfil a purpose within plants and trees, within cats and fleas, within everything? One does not know. One has to observe the world and try to make sense out of what one sees. The cat appears to be striving to catch the mouse; the plant appears to be striving to reach the Sun; but whether there are internal motivations to strive to fulfil those purposes ('catch the mouse'; 'reach the Sun') within these entities which are causing them to act in this way one cannot know for certain. To believe that a plant has an internal motivation to strive to 'reach the Sun' is not to believe that a plant thinks or has linguistic abilities; these internal motivations can be a simple attraction between two entities which is characterised by its qualitative feeling. So, there is a qualitative feeling in a plant which is attracted to the qualitative feeling of the Sun's energy. This attraction causes the plant to strive to reach the Sun – this is its purpose.

When I perceive the world around me the conclusion I reach is that there are internal motivations to strive to fulfil a purpose within everything; everything appears to be striving for something. Perhaps it is these internal motivations which 'make the world go around'; perhaps it is these internal motivations which are the reason life arose on the Earth; perhaps it is these internal motivations which caused the human species to come into existence; perhaps human beings have the internal motivations that they do so that they can fulfil a purpose. The more I think about this possibility the more compelling it seems; the human species has a purpose, planetary life has a purpose, everything is pervaded with purpose; everything is striving for something in an attempt to satisfy its own internal motivations.

Hopefully one agrees. If not, hopefully one can at least accept the possibility that life has a purpose, or that the human species has a purpose. If the human species is the only part of planetary life to have a purpose then the human species would obviously be special. However, even if, as I believe, everything has a purpose, then the human species could still be special. Is the human species special? Has planetary life been striving to evolve the human species and endowed it with a purpose of vital importance? If so, what is this purpose? This is the subject of the next chapter.

Chapter 4

Is the human species special?

What would it mean for the human species to be special? Clearly, if the universe had a creator and that creator endowed the human species with 'specialness' then the human species would be special. In theology the 'specialness' of the human species is referred to as the *Imago Dei*. The *Imago Dei* is the belief that in reality the human species is the only species of animal to have a 'special' relationship with the creator of the universe; the human species is created in the 'image' of this creator. What exactly is the *Imago Dei?* There have been numerous proposals as to the nature of this 'special' relationship.

In *Chapter Two* I proposed that the human species does not have any 'special' unique attributes. If I am right the source of human specialness (the *Imago Dei*) must lie elsewhere. If the universe did not have a creator this alternative source of human specialness could still exist. If there was no creator but planetary life had been striving to evolve the human species ever since it first arose, if the human species is the pinnacle of the evolutionary process, then the human species would be special.

In this chapter I propose that the human species is special because it is the pinnacle of the evolutionary process; it is the saviour of planetary life. If the universe had a creator and that creator particularly valued the living part of the universe, then one can see why there would be a special relationship between the human species and the creator. If the human species is the saviour of the planetary life-forms that the creator particularly valued then the human species would clearly have a special relationship with the creator. So, the view of the human species that I will be putting forward is also a proposal as to the possible nature of the *Imago Dei*. If the universe did not have a creator, the human species would still be special if it was the pinnacle of the evolutionary progression of planetary life.

According to one prominent contemporary view the rest of planetary life would be far better off if the human species were to become extinct! As the human species has spread out over the entire planet one effect has been that many species have become extinct, and unfortunately many more are in grave danger of extinction. The human species has turned natural habitats into 'concrete jungles'; it has initiated mass deforestation and mass agriculture; through its web of trade and transport links it has imported alien species into inappropriate habitats where the result has been the decimation of the native species; and it has released large amounts of oil into the oceans with disastrous effects. Now, many see the threat of human-induced global warming as the final 'nail in the coffin' of other species. According to this view, the human species through its selfish desire to plunder the world's resources in order to have a high standard of living is set to destroy a multitude of species, and possibly itself too.

So, if one has seen a lot of nature documentaries one will probably have heard many commentators assert that if the human species were to become extinct this would be great news for the rest of life on Earth. According to this currently popular view if the human species were to become extinct the rest of the species on the planet would be saved from our destructive influence. The advocates of this view believe that a couple of millennia after the extinction of the human species the biological diversity of the Earth would be vastly higher; life would supposedly be flourishing in the absence of the destructive humans. This view is grounded in what has happened in the past. In the past when there have been mass extinctions of life on Earth it *has* been the case that after a long period of time life has recovered; after a long enough period of time the biological diversity of the Earth has become just as rich as it was before the mass extinction. However, as I am sure you will appreciate, one cannot always use the past as a guide to the future.

My hope is to open you to the idea that this contemporarily fashionable view is wrong. The root of its wrongness is that it does not account for the fact that the Earth and the Sun are ageing entities. Furthermore, I contend that planetary life has been striving to bring the human species into existence since it first arose. So, if the human species were to become extinct now it would *not* be great news for planetary life; it would be an utter disaster.

Let us start with my main assumption. This is the assumption that *life wants to stay in existence.* It is widely accepted that life is a rare thing. The Earth is surrounded by a plethora of planets and stars and as far as we know the only life that currently exists in the universe is on the Earth; and, even if life exists elsewhere it is still a rare state for the universe. Life also seems to be a good state for the universe to be in. When one observes the Earth, either from the surface or from pictures taken from space, and then compares it to other planets there is something magical about it. A planet covered with life is a great thing compared to barren lifeless planets. On this, surely all can agree. Life appears to be a good state for the universe to be in. Given this, the possibility has to be seriously considered that when life arises on a planet it strives its hardest to stay in existence. When life arises on a planet an exceptionally rare event has occurred, a good state for the universe has been attained; the question is then whether this state can be maintained. Can life succeed in persisting? Or will its arising be a short term success, a temporary victory which is followed by the return to a barren planet?

There are plenty of reasons to believe that life on Earth has been very successfully striving to survive. When life first arose on the Earth it, presumably, only occupied a tiny fraction of the surface of the planet. Its continued existence was precarious. However, as the world we live in today is testimony to, planetary life was successful in its strivings. From simple beginnings planetary life successfully managed to spread out over the entire surface of the Earth. The forests are teeming with hundreds of thousands of species, the soil is full of life, the oceans and rivers contain a plethora of diverse life-forms; there is life in hydrothermal vents, life in the skies, life in the Antarctic and life in the sewers. Life has colonised the entire planet.

Is this surprising? Not if life strives to survive. However, if planetary life has no particular desire to stay in existence then it could have just limited itself to a small area of the planet, or it could simply have gone back out of existence soon after it first arose. The striving of life to survive reveals itself not only at the level of planetary life; it also reveals itself at the level of a single life-form. The human body strives to stay alive as long as possible; when it is cut it tries its hardest to stop the bleeding; when it is infected it tries its hardest to fight the infection. The human body, like life as a whole, has no desire for death, for extermination.

There is a chance that you might be thinking: *this must be wrong – if the human body has no desire for death why do some humans commit suicide.* So, I will explain what I mean in a little more detail. I suggest that the human body – at the level of atoms, cells, and organs – is pervaded with a desire to survive, a desire to maintain a state of life. I am proposing that this desire exists through-out planetary life. In addition to this desire some parts of planetary life also have the capacity for thought. I propose that thought is wholly devoid of desire, wholly devoid of striving, and wholly devoid of the qualitative feeling which exists in a state of desire. On this view the human brain can be thought of as having two elements. Firstly, it is pervaded with desire because it is part of the body. Secondly, the interactions of these desires enable it to contain thought at a 'higher-level' than these desires. Suicide occurs when 'higher-level' thought overpowers the desire of the body to maintain a state of life.

Why did planetary life strive to colonise the entire planet? One reason is strength in numbers. The more life-forms that exist in a particular location, and the more locations that life-forms exist in, the more likely it is that some life-forms will survive. This is because all life-forms require particular conditions in order to survive and the conditions in a particular part of the planet often change. So, if a particular ecosystem contains a wide-range of life-forms, each of which requires slightly different conditions in order to survive, then many of these life-forms will survive if the ecosystem conditions change (e.g. a higher temperature, a reduction in nutrients, or a change in atmospheric composition). In ecological terminology such an ecosystem has a high 'redundancy'.

There is also a more complex reason for the colonisation of the entire planet. The foundation of this reason is that *life needs a specific set of conditions in order to survive but the conditions of the universe are always in flux.* By 'conditions of the universe' I do not mean conditions within a particular ecosystem; I mean changes that gradually affect the entire biosphere of the planet over timescales of tens of thousands of years. By colonising the entire planet planetary life was able to start manipulating and subtly affecting the conditions of the planet in order to keep them favourable for its continued existence. This phenomenon is typically referred to as the Gaia Theory. This term was coined by Sir James Lovelock who was the first person to recognise the existence of this phenomenon. Today those who study the mechanisms underpinning Gaia Theory are typically referred to as Earth Systems Scientists.

The best way to appreciate the Gaia Theory is to understand that since life arose on the Earth the amount of sunlight which reaches the Earth has increased dramatically. Of course, it is energy from the Sun which enabled life to arise in the first place and which still enables life to survive. The important fact is that, since life arose, slowly but surely the Sun has been getting hotter and hotter which means that the amount of energy which reaches the Earth from the Sun has increased dramatically. The Sun's energy output has increased by 40% since it was formed and 25% of this increase has occurred since life arose on the Earth. Initially, soon after life arose, this increase in energy was good for life. But, life needs a specific set of conditions to survive, and complex life (such as humans) needs a very specific set of conditions; planetary life, and complex life in particular, cannot survive if the atmosphere of the planet is too hot. If planetary life was to complexify and flourish it needed to regulate the planetary conditions in order to keep the atmospheric temperature steady despite a forever increasing amount of energy coming in from the Sun. And this is exactly what planetary life did.

From the Gaia Theory perspective all of the fossil fuels that are stored in the Earth's crust are stores of carbon which were deposited there in order to take carbon dioxide out of the atmosphere. Less carbon dioxide in the atmosphere means that the temperature of the atmosphere can be sustained despite increasing energy coming into the planet from the Sun. Planetary life has done a great job in maintaining the conditions needed for life to complexify and stay in existence despite the changing conditions caused by the Sun. However, there are limits to the ability of planetary life to regulate the temperature of the Earth's atmosphere. The Sun, who is the provider of life, is now becoming a formidable adversary to life. Life's ability to regulate the atmospheric temperature is weakening as the Sun's energy relentlessly keeps on increasing.

Why is the ability of planetary life to regulate the temperature of the Earth's atmosphere weakening? Planetary life has been keeping the temperature of the Earth's atmosphere down by taking carbon dioxide out of the atmosphere. In this way the increasing energy from the Sun has been offset and the Earth's atmospheric temperature has been successfully regulated. However, there is only so much carbon dioxide in the atmosphere; there is a point at which the increasing energy from the Sun can no longer be offset in this way. As James Lovelock has realised, the planetary fluctuations between ice ages and interglacial periods are an indicator that the ability of planetary life to regulate the temperature of the Earth's atmosphere is weakening. At the start of ice ages planetary life is fairly comfortably regulating the temperature of the Earth's atmosphere. However, as time progresses within an ice age the amount of carbon dioxide in the atmosphere falls which makes it much more difficult for planetary life to regulate the atmospheric temperature; the less carbon dioxide that there is in the atmosphere the harder it is to take carbon dioxide out of the atmosphere. This increasing 'difficulty' leads to an interglacial period – a period of 'stressful' regulation. The interglacial period ends – giving planetary life a temporary respite – when the tilt of the

Earth changes slightly which temporarily reduces the amount of energy which reaches the Earth from the Sun.

In the past, before there were fluctuations between ice ages and interglacial periods, planetary life was easily able to regulate the temperature of the Earth's atmosphere. The fluctuations are an indicator that all is not currently well. From this perspective the future looks bleak; the current weakening of the ability of planetary life to regulate the Earth's atmospheric temperature looks set to be followed by a *total collapse*. This is because the point will come when planetary life is unable to offset the increasing energy from the Sun because there is not enough carbon dioxide left in the atmosphere. Human actions have, so far, increased the likelihood that this total collapse will occur in the near future. In a short period of time the human species has released a colossal amount of the previously stored fossil fuels into the atmosphere thereby contributing to the forces which are warming up the atmosphere of the Earth. Whilst, simultaneously, the human species has initiated mass deforestation thereby significantly reducing the ability of planetary life to keep the temperature of the atmosphere down (forests play a key role in regulating the amount of carbon dioxide in the atmosphere). So, there are two routes to a total collapse. Firstly, a slowly reducing level of carbon dioxide in the atmosphere eventually leads to a total collapse. Secondly, a short-term release of colossal amounts of carbon into the atmosphere when combined with mass deforestation can lead to a more immediate total collapse.

Perhaps this is a good point to reflect on the question of why life, which had simple beginnings, not only colonised the entire planet, but also complexified into forms as complex as humans. One obvious possibility is that the more complex a life-form is the more ability it is likely to have to be able to modify its surroundings. Another related possibility is that the striving for complexity is a striving for one particular species; a species that can become the modifier *par-excellence;* a species that can analyse in detail its surroundings and modify them in extraordinary ways. In short, planetary life could have complexified in order to evolve a technological species.

What is a technological species? We saw in *Chapter Two* that numerous species of animals are tool-users. Tool use is a central part of being technological, but there is much more to being technological than just tool-use. A technological species is able to design tools of immense complexity – tools such as telescopes, mobile phones, satellites, space stations and televisions. The design of tools of such complexity requires in-depth knowledge of the universe. So, a technological species is a species which uses in-depth knowledge of the universe to create tools of immense complexity.

Why would planetary life need to evolve a technological species in order to fulfil its desire to survive? As we have already seen, the Sun is becoming a formidable adversary to planetary life and the ability of planetary life to maintain the temperature of the Earth's atmosphere is inevitably weakening. Well, perhaps this is not quite right. One should actually say, *in the absence of the human species* the ability of planetary life to maintain the temperature of the Earth's atmosphere is inevitably weakening. Without the human species in the not too distant future the ability of planetary life to maintain an atmospheric temperature favourable for complex life-forms will have disappeared. But all is not lost for planetary life because it has complexified into a technological species – the human species – which is fast accruing the ability to take over the helm by developing the technology which enables the temperature of the Earth's atmosphere to be regulated, to be kept down, in the face of the forever increasing output of the Sun. It would be disastrous for planetary life if the human species were to become extinct because bringing forth a technological species is a very slow process and planetary life needs technology in the near future; there wouldn't be enough time to bring forth another technological species.

Why would the human species develop such technology? In fact, many such technologies are already being designed; technologies such as an array of satellites which would deflect the rays of the Sun; when I talk of 'technology which regulates the temperature of the Earth's atmosphere' such an array of satellites is what I have in mind. Do these designers realise that the human species has a purpose? Do they realise that they are playing a major part in the purpose of the human species; that they are fulfilling the objectives of life; that they are being the saviours of planetary life? This seems to be very unlikely. They probably believe that they are simply developing technology to deal with human-induced global warming.

Could it be that in order to fulfil its purpose and be the saviour of planetary life that humanity had to believe that it was potentially the destroyer of planetary life? Possibly. But I don't mean to suggest that human-induced global warming is a mirage. It is obviously a real phenomenon, a real threat. I am simply suggesting that in order to fulfil the purpose for which it came into existence – to save planetary life – it might have been necessary for the human species to warm up the planet a little. Why else would humans have good reason to develop the much needed technology? There are, no doubt, many people who think that human technology should not be used to attempt to control the temperature of the Earth's atmosphere because such an endeavour would be too risky. I can only hope that on reading this book these people come to believe that the only way planetary life has any chance of survival is through the development and deployment of such technology. Without it planetary life is doomed. I am confident that in the near future the human species will be regulating the temperature of the Earth's atmosphere. This seems to be the destiny that planetary life has been heading towards for millions of years.

One might be wondering what would have happened if the human species had stayed living in small communities and had developed only simple tools. Without the scientific and industrial revolutions the human actions which have given rise to the fears about human-induced global warming would never have occurred. In this scenario the needed technology would not have been developed. I reply that this scenario could never have transpired. The human species by its very nature, its internal motivations, was always going to become technological (the internal motivations of humans are explored in *Chapter Six*). In short, cultural evolution has a rough trajectory from hunter-gatherer to globalised technological society. The impetus behind this trajectory could not have been halted. The processes at work which give rise to this trajectory will be explored in later chapters.

So, our conclusion is that the human species is special. Its purpose is the most important purpose there could be. The purpose of the human species is to be the saviour of planetary life by developing and deploying the technology that will maintain the temperature of the Earth's atmosphere as the non-technological ability of life to offset the increasing output of the Sun weakens.

Chapter 5

Reflections on the environmental crisis

When reflecting on the changes that have been made to the Earth by the human species it is useful to distinguish between three phenomena – the environmental crisis, human-induced global warming and human-induced climate change. Let us consider the phenomenon of human-induced global warming first.

Humans, through a variety of activities – deforestation, farming, use of fossil fuels – have increased the amount of carbon dioxide and methane that exists in the Earth's atmosphere. Carbon dioxide and methane trap the infrared radiation from the Sun after it has bounced off the surface of the Earth and are therefore called greenhouse gases; the result is a higher planetary temperature than would otherwise have been the case. The phenomenon of human-induced global warming refers solely to this fact – the activities of the human species causes the temperature of the Earth's atmosphere to be higher than it would have been if the human species had never existed.

Is human-induced global warming a cause for great concern? This is a question which is vigorously debated.

There are those who believe that human activities will only have a very small impact on the temperature of the Earth's atmosphere; according to these people human-induced global warming is not a cause for great concern. However, there are far more people who believe that the impact on the temperature of the Earth's atmosphere could be very large. These people are concerned, and quite rightly so, about a number of very worrying possibilities. It is possible that the Earth's atmosphere, rather than changing gradually, could suddenly flip from one state to another utterly different state when it passes a certain threshold. This can be thought of as the 'tipping point' scenario. Another worry is that the Earth's atmosphere might initially change gradually and then change in an accelerated manner – temperatures could shoot upwards rapidly. This is the 'runaway greenhouse effect' scenario.

These possibilities are made very real because of the likelihood that a lot of the carbon which the human species has released has become stored up waiting to explode with one massive effect. This is analogous to when one pulls an elastic band further and further back and then eventually releases it. One significant storage area is the oceanic thermohaline circulation. In the Polar Regions the density of surface water increases causing it to sink deep into the ocean where it moves very slowly until it eventually reappears in other regions such as the North Pacific; these deep ocean currents take 1600 years to move from the

North Atlantic to the North Pacific. Increasing amounts of human-released carbon have become stored in these deep ocean currents – this increasing storage is the 'pulling of the elastic band'. The 'release of the elastic band' will occur when these currents reach the surface and the carbon dioxide starts to be released into the atmosphere; this could lead to either a 'tipping point' scenario or a 'runaway greenhouse effect' scenario.

Given our ignorance about the possible effects of our activities on the Earth's temperature it has to be sensible to adopt the 'precautionary principle' – to act as though the worst case scenario could happen. If we don't and the worst case scenario does happen then the Earth could switch to a state in which no complex life-forms are able to survive.

Let us now consider the phenomenon of human-induced climate change. This phenomenon is linked to global warming because if the temperature of the Earth's atmosphere rises, one effect of this is to alter the climate in various parts of the world. However, the phenomenon of climate change itself is simply *a long-term change in weather patterns*. Climate change occurs when areas of the Earth that have had a regular pattern of plentiful rainfall for decades change to a state in which there is no rainfall for decades. Climate change also occurs when areas of the Earth that have had consistently hot temperatures for centuries suddenly encounter cold conditions for centuries. When humans engage in deforestation they can alter the climate in specific parts of the world. However, if they also plant an 'offsetting' amount of trees somewhere else on the planet there will be no human-induced global warming resulting from this deforestation/climate change. There can be climate change without global warming; and there can be human-induced climate change without human-induced global warming.

Is human-induced climate change a cause for great concern? It is a cause for concern but it isn't as worrying as global warming. There are two main issues surrounding climate change – the way that the human species has organised itself into nation states *and* the current state of development across the planet. In the past when the climate changed humans were simply able to move to parts of the planet which were more hospitable. However, now that the surface of the planet has been carved up into nation states, this movement is more problematic. The main cause for concern raised by climate change arises from the differential state of development across the planet. 'Developed' countries have enough resources to be able to adapt to moderate climate change. However, climate change could be much more worrying for those humans, and other life-forms, which live in 'developing' countries.

Let us now consider the phenomenon of the environmental crisis. This is a *purely* human-induced phenomenon (climate change and global warming can both occur in the *absence* of humans). The human species affects its environment just by existing. Of course, merely affecting the environment does not entail an environmental crisis. The reality is that the human species has affected its environment in *a number of very large ways* – ways that affect the entire planet and which are very long-lasting. It is this family of very large and very worrying changes which humans have made that *as a whole* constitute the environmental crisis. This family of changes includes the mass extinction of other life-forms, climate change, global warming and the weakening of the ozone layer. The environmental crisis is a term which refers both to changes which have already occurred and to fears about likely changes which look set to occur in the future. The environmental crisis is obviously very concerning.

Now that the difference between these three phenomena has been outlined let us turn our attention elsewhere. In the previous chapter I proposed that since life arose on Earth it has been *striving to evolve a technological species*. A species becomes technological when it goes through a scientific revolution and an industrial revolution. Scientific enquiry provides the detailed knowledge which enables a species to modify its surroundings in intricate ways through industry; the technological tools which are part of industry (e.g. complex machinery in factories, steam engines, MRI scanning equipment) far exceed non-technological tool-use; technological tools turn a tool-using species into a technological species. In *Chapter Two* I made a distinction between 'non-advanced tools' and 'advanced tools'. The category of 'advanced tools' includes both 'technological tools' and 'non-technological tools'.

When a species first becomes technological it does not realise its power. At first industry uses a minimal amount of resources. But as time passes industry inevitably uses up more and more of its planet's limited resources. This effect is reinforced by another inevitable phenomenon; a technological species becomes a more healthy and long-living species. Scientific knowledge in tandem with the pharmaceutical industry and the creation of complex medical equipment leads to an increase in life-span and an increase in population size; this knowledge and technology enables many individuals to survive who would otherwise have died. The industrial revolution led to a dramatic increase in the life expectancy of children. There are other factors which also cause the population size of a techno-logical species to vastly increase. Technology can be used to transform areas of the planet into more favourable habitats for the species (e.g. advanced irrigation systems can supply water to arid areas). Technology can also be used to hugely increase the amount of food that is available which results in an escalating population. The 'Green Revolution' is a prime example; technology enabled the development of high-yielding varieties of grains, hybridised seeds and synthetic fertilisers which were used in 'developing' countries and resulted in a rapid increase in the size of the population.

The beginnings of the Industrial Revolution were around the mid-18th century. At the beginning of the 19th century the human population was approximately 1 billion; today the human population is approaching 7 billion. When a species becomes technological its population explodes.

The combination of this population explosion with the ability of industry to access and use increasing amounts of the Earth's resources leads to an inevitable outcome – a planetary environmental crisis. In other words, it is inevitable that a technological species will initiate a planetary environmental crisis. At a certain time a technological species comes to realise how powerful it is; comes to realise how much damage it has done to the environment of its planet. The trouble is that by the time a technological species realises this, its population is so large in size and has such ingrained habits that it cannot easily stop its previous course.

Recall that in the previous chapter I proposed that since life arose on Earth it has been *striving to evolve a technological species.* Now we can see that, in effect, I am proposing that since life arose on Earth it has been *striving to initiate a technology-induced planetary environmental crisis.* This will, perhaps, sound like a hard idea to accept: *why would it be in the interests of life for there to be a planetary environmental crisis?* The answer is that an environmental crisis is in the interests of life if it is the only way it can evolve a technological species; planetary life wants to survive and in order to survive for much longer it needs a technological species. I am suggesting that life cannot develop a technological species without that species initiating an environmental crisis. Technology inevitably leads to an environmental crisis and part of this environmental crisis is *human-induced global warming.* This is in the interests of life *if* in order to combat human-induced global warming the human species develops and deploys the technology which regulates the temperature of the Earth's atmosphere. I suggest that there are forces in existence which will ensure that this deployment occurs. If this is right, one can see why life strives for an environmental crisis; it is the only way it can get *what it really wants* which is that part of the environmental crisis which is human-induced global warming *and* the human technological response. So, the environmental crisis is a great event for life; a small short-term price to pay for its long-term survival.

I wouldn't want you to get the wrong idea about what I am saying here. I wouldn't want you to think: *if an environmental crisis is such a good thing I will stop recycling, take more flights and drive my car more!* If one is environmentally aware, that is fantastic. To the environmentally aware person I say: *recycle more, fly less and drive less!* Environmental awareness is a great thing. If it were not for environmentally aware people there would not be any pressure on governments to deal with human-induced global warming (which ultimately means they will need to develop and deploy the technology that planetary life so badly needs; the technology that can regulate the temperature of the Earth's atmosphere). Being environmentally aware and acting in an environmentally friendly manner is also a good thing in its own right. When some humans realise what their species has done to the planet they will seek to live in a more environmentally friendly manner. Some of these humans will also try to help other members of their species to live in a more environmentally friendly manner. These individuals are doing great work and should be commended for their work. I have particular admiration for those who campaign against vivisection, factory farming and all types of animal cruelty. Those who research and campaign on human-induced global warming are also doing very valuable work.

What I am saying is that it is when one takes a very broad view which includes the formation of the planet Earth, the origin of planetary life, the origin of the human species, and all of human history, that the environmental crisis can be seen to be a 'good' thing. This conclusion has no bearing on how one should act if one is a human living at a time of growing environmental awareness; that is to say, there is no generally applicable way in which everyone should act.

Do I appear to be contradicting myself? On the one hand the environmental crisis is a 'good' thing, on the other hand people living in a more environmentally friendly manner is also a good thing. How can this be? Let me explain. One needs to imagine that there are two forces at work in global culture. The overwhelmingly dominant force is the force that has been inevitably increasing in strength for thousands of years; it is a paradigm example of the 'snowball effect'; this force is *the force to environmental destruction.* This force is powered by an escalating global human population and the amount of resources that are used by the individuals in that population. This force is currently so powerful for a number of reasons. Firstly, when a species becomes technological it is inevitable that only part of the species becomes technological first. The scientific and industrial revolutions had to happen somewhere in particular. As it turned out it was in Western Europe that the human species became technological. Humans in this part of the world used this to their advantage and dramatically increased their living standards/resource use per head. This increase in living standards was inevitable. Secondly, the escalation in population that occurs when a species becomes technological is also inevitable. These two factors – an escalating population, with a significant proportion of that population with a high resource use per head, is a potent force for environmental destruction. This force is reinforced by a third factor. The majority of people who have experienced living with a high resource use per head have ingrained habits/ways of living, and are typically very reluctant to any suggestion that they should use far less resources; it would almost seem reasonable to suggest that the majority are *incapable* of making anything other than piecemeal changes to the way they live. Fourthly, this whole situation is made woefully worse by the fact that both the new members who are born into the escalating population, and the existing members of the population who are living on a very low resource use per head (as in large parts of Africa), have a large proportion of members who, quite understandably, aspire to reach the living standards of those with a high resource use per head. This combination of forces is an almost unstoppable juggernaut heading for environmental destruction.

Let me say a bit more about the habits of those who have a high resource use per head and of those who live in the 'developed' countries more generally. There are, no doubt, a tiny minority of people in a 'developed' country who live in a truly sustainable manner. Such people might have no car, never take any aeroplane flights, and might grow all their own food. Good for them. Let us consider the overwhelming majority of people (which includes the vast majority of people who are 'environmentally aware'). It is astounding when one considers how many resources a typical person in a 'developed' country uses in a year. A typical person in one of these countries will probably have a lifestyle which includes most of the following: a car or motorcycle which is used almost daily, several aeroplane flights a year, a diet with a large proportion of items which have been flown from distant countries, a house or flat full of hardly used items and unwanted gifts, a habit of buying things they don't need, numerous electrical appliances, a desire for lots of holidays in exotic locations, and a bin full of unnecessary packaging and junk mail/magazines/newspapers. The unsustainable habits of these hundreds of millions of people can be changed, but only in piecemeal ways (recycling bottles/cardboard/tins; taking bags to the supermarket). After these piecemeal changes are taken into account their overall way of living is still *highly* unsustainable and is not something which is going to be given up because of talk of an environmental crisis. It is better to be realistic about the direness and the immediacy of the situation than to expect hundreds of millions of people to suddenly change *en mass*.

The second force at work in global culture is the *force to environmental sustainability*. This force is in its infancy, it is a force which has only had any real power since the second half of the twentieth century. This force is at work when some people feel slightly guilty about their unsustainable lifestyles. It is in operation when some people take an aeroplane flight to the other side of the world to take an 'eco-holiday' rather than a normal holiday. This force includes much of the recycling that goes on, the reuse of materials, and the development of wind farms and other renewable energy sources. This force is also in operation when people campaign against deforestation and human-induced global warming, and when political leaders meet up to discuss these issues.

Both of these forces are inevitable components of global culture at this moment in time. At the moment the *force to environmental destruction* is overwhelmingly dominant. This is why when political leaders meet to discuss human-induced global warming they are unable to make the strong commitments that those who are part of the *force to environmental sustainability* urge them to make. This outcome should not surprise us. I am also suggesting that it should not worry us. The *force to environmental sustainability* is a very valuable force and its strength will inevitably continue to grow. This will be a good thing; the renewable energy technologies which are slowly being developed will be of vital importance in the future. But the state of the world at the moment is that the *force to environmental destruction* dominates. I propose that this is the way things should be. It is in the interests of life that the *force to environmental destruction* should, at this moment in the evolution of human culture, still continue to dominate.

What would happen if the *force to environmental sustainability,* suddenly, miraculously, became over-whelmingly dominant? Let us consider two scenarios. In the first scenario all of the humans in 'developed' countries immediately change their lifestyles so that they stop using electricity for televisions, music systems, DVD players and computers, stop driving their cars forever, use a minimal amount of resources in all areas of their life, and commit to never take an aeroplane again? And, in tandem, all of the humans in 'developing' countries decide to continue to live on a very low resource use per head. In the second scenario there is a sudden explosion and proliferation of renewable technologies around the world which enables the humans in 'developed' countries to maintain their lifestyles whilst having negligible environmental impact; whilst the humans in 'developing' countries are able to significantly raise their resource use per head without environmentally deleterious effects.

If either of these scenarios pertained then there would be an *instantaneous* global victory for environmental sustainability. This would mean that the outspoken advocates of the *force to environmental sustainability* would consider their objectives achieved, and human-induced global warming would become a far less discussed issue. Indeed, many would assume that human-induced global warming was no longer a problem, and these people might be right. So, in this scenario it is entirely possible that the atmospheric temperature-controlling technology that planetary life (which obviously includes the human species) needs so desperately in order to survive would not be developed.

Would this be a good thing? No! If the *force to environmental sustainability* suddenly became all-powerful this would potentially be disastrous for planetary life. So, I am suggesting that things are as they should be. It is in the interests of life that the *force to environmental destruction* continues to dominate. This force, this inevitable juggernaut, is going to leave the human species with no choice but to develop the technology that life so badly needs; the technology which will maintain the temperature of the Earth's atmosphere despite the forever increasing output of the Sun. In short, the continued dominance of the *force to environmental destruction* will ensure that there is a small amount of human-induced global warming. The effects of this warming and worries about the future will cause the human species to utilise all of its technological expertise in order to regulate the temperature of the planet's atmosphere. I am suggesting that this is inevitable. It is the purpose for which the human species came into existence.

The effects of human-induced global warming – which are the stimulus for the development of the required technology – are themselves wholly undesirable. Human-induced global warming could result in many low-lying islands becoming submerged as sea-levels rise; human-induced climate change resulting from human-induced global warming could result in droughts and famines; many species could see their habitats disappear resulting in their extinction. The possible deleterious effects are almost too numerous to mention. *Thankfully the human species is able to do something about this situation.* What can be done to minimise these effects? If the picture I have been presenting is correct there is only one answer to this question. The *only* way that the deleterious effects of human-induced global warming can be *minimised* is to develop and deploy the technology which regulates the temperature of the Earth's atmosphere as quickly as possible. This development and deployment is inevitable, but if it is done speedily the deleterious effects of human-induced global warming can be minimised; indeed human-induced global warming itself can be halted. I should say that by 'inevitable' I mean will occur in the future *unless* a 'freak event' occurs such as the human species being wiped out due to the collision of a massive asteroid with the Earth.

If the human species were *solely* to engage in 'traditional' responses to human-induced global warming – reuse of resources, recycling, building sea defences, renewable energy technologies – the danger is that this could lead to unnecessary suffering and a plethora of other avoidable deleterious effects. These measures themselves are all very desirable but they are not capable of offsetting the *force to environmental destruction*. This destructive force is *overwhelmingly powerful* and it is *still* increasing greatly in strength; this needs to be fully acknowledged. One should not be deceived by the fairly widespread coverage of environmental issues in the news, and piecemeal changes such as the increase in recycling and the deployment of wind turbines; the fact is that the *force to environmental destruction* is still getting stronger and stronger; it dwarfs the *force to environmental sustainability*. All elements of the *force to environmental sustainability* are useful but the battle will inevitably be lost unless the *force to environmental sustainability* has its trump card in play – the development and deployment, by the human species, of the technology which regulates the atmospheric temperature of the entire planet.

Let me be clear. It is true that the environmental crisis and human-induced global warming *are* real threats to planetary life. However, there is a vastly larger threat to planetary life – a threat which was not caused by the human species, but a threat which the human species can neutralise. Both of these threats are real and both exist *right now*. I believe that we have ample time to fulfil our purpose as a species; ample time to develop and deploy the required technology. But, as we have seen, we cannot know how the systems of the biosphere will change in the future. A tipping point could be passed or there could be a runaway greenhouse effect, and *these eventualities could be catalysed by either human actions or by non-human factors such as the increasing output of the Sun.* The future is most uncertain. The most sensible course of action is to fulfil our purpose as a species as soon as possible. This would deal with both human-induced global warming and with the vastly larger non-human threat to the continued existence of planetary life.

It is possible that one might still be convinced that it is human-induced global warming that is the major threat that needs to be addressed. So, let me repeat what I have said in a slightly different way. Let us assume that the effects of the human species on the temperature of the Earth are neutral. In other words, there is no human-induced global warming now and there will be none in the future. In this scenario, it would, of course, still be the case that the vastly larger threat of global warming to planetary life still remains and needs to be dealt with very soon. And, the only way that planetary life can be saved from this threat is by the human species developing and deploying the required technology; if we fail all of the wonderful life-forms that currently exist on the Earth will be decimated.

I envision that when this technology is in place the *force to environmental sustainability* will be much stronger. Indeed, it is easy to imagine that with this technology in place, perhaps as an array of satellites in the sky reflecting the rays of the Sun, that people will become much more aware of the effects of their actions. The array of satellites would be a potent symbol. The result could be that the two forces – sustainability and destruction – become balanced. The cancelling out of these forces would mean that not only is the planetary temperature stabilised, but that environmental sustainability would exist across the planet in all areas. No more deforestation, no more over-exploitation of the seas, respect for all life-forms. One can hope that this could be the future. I fear that it is not attainable until our purpose as a species has been fulfilled.

Chapter 6

Do individual humans have a purpose?

I have been presenting a view according to which the human species has a very special purpose. The purpose of the human species is to develop and deploy the technology that will regulate the temperature of the Earth's atmosphere. Being endowed with this purpose means that the human species is the saviour of planetary life. Without this technology the ability of planetary life to regulate the temperature of the Earth's atmosphere will continue to weaken and there will be devastation; the overwhelming majority of the wonderful life-forms that currently exist will no longer be able to survive and the Earth will, at best, only be able to sustain simple life-forms.

You may be wondering if in addition to the human species having a purpose, individual humans have a purpose too. By this I mean does *every* human have a purpose. Does every human have a role to play in achieving the purpose of the species? Obviously it would be the case that those particular humans who are involved with the development, deployment and maintenance of the required technology are fulfilling a purpose. However, these humans are fulfilling the purpose of the *species*. Was this *also* their unique purpose as individual humans?

If individual humans do *not* have a unique purpose to play in fulfilling the purpose of the species then one might wonder how it is possible for the purpose of the species to be fulfilled. Let us consider an analogy between the human species and a football team. The purpose of a football team is to be victorious over their opposition. One can understand how this purpose can be achieved because each of the members of the team has a unique individual purpose within the team. The goalkeeper has the purpose of stopping the other team from scoring, the defenders each have the purpose of giving protection to the goalkeeper in a particular area of the pitch, and the striker has the purpose of scoring goals. Because of these individual purposes it is easy to grasp how the purpose of the team can be fulfilled. If all members of the team fulfil their individual purpose then the purpose of the team can be fulfilled. However, if one considers the possibility that the team has the purpose of being victorious, but that the individual members have no individual purpose, then it is much harder to imagine how the purpose of the team can be fulfilled. If the goalkeeper does not have the purpose of stopping the other team scoring he could simply stand still as the opposing striker kicks the ball towards the goal. If the striker does not have the purpose of scoring goals he might simply decide to take a nap on the pitch.

Of course, it is possible that such a team might be victorious. But if it was, one would be rather puzzled as to how such an extremely unlikely outcome occurred. For this reason we need to consider the relationship between the purpose of the human species and the actions of the individual members of that species. Do individual humans have purposes the summation of which results in the purpose of the species as a whole being fulfilled?

How can one make sense of the idea that every human has a purpose? Let us consider two possibilities. Firstly, one could envision that in order for the purpose of the species to be fulfilled that there are a massive number of 'mini purposes' which need to be fulfilled. For example, the wheel needed to be invented. The invention of the wheel is a 'mini purpose' which plays a part in fulfilling the purpose of the species. One can envision that 'mini purposes' exist prior to the individuals who in actuality fulfil that purpose; the 'mini purpose' is an 'empty vessel' waiting to be filled. There is nothing mysterious about these 'mini purposes'/'empty vessels'. We are assuming that from the moment that the human species first evolved it had a purpose as a species which needed to be fulfilled. From this first moment, this beginning of the species, all of the steps which needed to be taken in order for that purpose to be fulfilled were already clear. One such step is 'the invention of the wheel', another is 'the initiation of the scientific revolution', another is 'the invention of the telescope', and another is 'the recognition of human-induced global warming '. These required steps are 'empty vessels' which need to be filled; in filling these vessels humans are fulfilling 'mini purposes'. When all of the 'mini purposes' are fulfilled the human species will have fulfilled its purpose as a species.

On this 'empty vessel' view it doesn't matter which individual human fills a vessel. As long as some human fills an empty vessel the species will be another step closer to the fulfilment of its purpose. So, on this view there isn't one particular individual human whose purpose it was to invent the wheel. Nevertheless, there was an individual human who invented the wheel and thereby fulfilled a 'mini purpose' which contributed to the fulfilling of the purpose of the species. It is important to stress that it is also possible for lots of people to fulfil the same 'mini purpose'; one example is 'use an unsustainable amount of resources'.

Of course, one might consider it to be a possibility that no member of the human species ever invented the wheel. If so, then one might find all this talk of the 'empty vessel' of 'inventing the wheel' to be slightly odd. However, I hope that when one reflects on the ascent of the human species from humble beginnings to astronaut one will agree with me that it is highly implausible to assert that the human species would not, at some stage, have invented the wheel. This means that from the very beginnings of the human species *it does* make sense to talk of the 'empty vessel' of 'inventing the wheel' which was waiting to be filled.

The 'empty vessel' view is quite appealing because one will probably find it very hard to accept that there was one individual human who came into existence for the specific purpose of inventing the wheel. What would have happened if this particular individual was the unfortunate subject of a terrible accident early in their life? Would the wheel never have been invented? One surely wants to accept that things could have been different in the past – *the actual person who invented the wheel could have died at a young age and if they did surely someone else would then have invented the wheel, so it cannot have been the case that it was the unique purpose of the actual person who invented the wheel to be the inventor of the wheel.*

The alternative to the 'empty vessel' view is the view that it was the unique purpose of the actual inventor of the wheel to be the inventor of the wheel. As we have seen the problem with this view is that if this particular human for some reason did not fulfil their purpose – perhaps they had an untimely death, perhaps they were very unfortunate and succumbed to a debilitating disease, and perhaps they just didn't want to fulfil their purpose, then one still wants to think that the wheel would have been invented. Of course, this is only a problem for the view on the assumption that it was the unique purpose of the inventor of the wheel to be the inventor of the wheel from the moment of their birth. However, if it was only their unique purpose at the moment when they actually fulfilled the purpose, then this view seems indistinguishable from the 'empty vessel' view.

In the remainder of the chapter I present an account according to which individual humans have a purpose. In this account the vast majority of humans play a part in the fulfilment of the purpose of the human species. The account I will be outlining is a type of 'empty vessel' view of human purpose as this view seems to be the most plausible view of individual human purpose.

So, let us consider how individual humans come to fulfil all of the 'mini purposes' that need to be fulfilled if the species is to fulfil its purpose. How do individual humans come to fill all of those 'empty vessels'? I have used 'inventing the wheel' as an example of a 'mini purpose'. There are an immense number of 'mini purposes' which need to be fulfilled if the human species is to fulfil its purpose. And, of course, many of these 'mini purposes' have no chance of being fulfilled until *other* 'mini purposes' are fulfilled. One cannot fulfil the 'mini purpose' of inventing the bicycle until the 'mini purpose' of inventing the wheel has been fulfilled. This is why cultural evolution can only progress at a certain speed. Sometimes the fulfilment of 'mini purposes' is an exceptionally slow process, whilst at other times a flurry of 'mini purposes' seem to be fulfilled in a very short period of time. The early twenty-first century seems to be a period in which a plethora of 'mini purposes' are getting fulfilled relatively quickly. When old people alive today look back to when they were young, and contemplate the evolution of human culture more generally, they often claim: *things seem to be speeding up; today everything changes much faster than it did in the past.* Perhaps they are right.

By what mechanism do particular 'mini purposes' come to be fulfilled by particular humans? I suggest that there is a very subtle mechanism in place which ensures that 'mini purposes' get fulfilled. This mechanism is centered on human potential. Let us consider how one would judge as to whether or not a human has fulfilled their potential. In so judging one does not use as the benchmark of fulfilment a 'mini purpose' such as 'inventing the wheel'. Many humans fulfil their potential without fulfilling such an important 'mini purpose', whilst others would not be fulfilling their potential solely by fulfilling this one 'mini purpose'. Human potential varies greatly and is something internal to a specific human.

So, when one considers whether a particular human is fulfilling their potential one considers factors such as the following: Has this human tried hard to utilise all of their talents and abilities? Has this human embraced life and sought to do that which makes them satisfied/content? Does this human find their work exhilarating? Is this human being all that they can be? Does this human have a sparkle in their eye? If at least a couple of these questions have positive answers then it is probably fair to say that this human is fulfilling their potential. So, fulfilling one's potential requires knowing oneself; one needs to know what one's talents and abilities are if one is to become all that one can be, if one is to be satisfied and content.

What does fulfilling one's potential have to do with fulfilling a 'mini purpose'? I suggest that if one fulfils one's potential then it is inevitable that one will fulfil at least one 'mini purpose'. There are, no doubt, some humans who go their entire life without fulfilling a 'mini purpose'. The reason that these humans do not fulfil a 'mini purpose' is that they do not fulfil their potential. Now, for the human species as a whole to fulfil its purpose it has to be the case that the vast majority of humans fulfil 'mini purposes'. If only a minority of humans fulfilled a 'mini purpose' then the majority would be in control and the result of this would be that human culture would take a different trajectory to the one which is required if the human species is to fulfil its purpose.

How can the vast majority of humans be relied upon to fulfil their individual potentials and thereby ensure that the human species as a whole fulfils its purpose? The majority can be relied upon because fulfilling one's potential is the natural state of affairs. There are forces in existence which ensure that most people fulfil their potential – forces such as happiness, contentment, bliss, satisfaction, financial reward, peer pressure, the attraction of a partner and a healthy and long life. Those humans who do not fulfil their potential are *more likely* to be unsatisfied, miserable and poor. It is, of course, possible to both fulfil one's potential *and* be poor; however, financial reward clearly helps to motivate *some* people to fulfil their potential.

So, we can see why the majority of humans are likely to fulfil their potential; most want to be satisfied, healthy, and so on. And we can also understand why some humans will not fulfil their potential. Some humans develop negative ways of thinking, get stuck in deleterious habits, and lack the confidence to fulfil their potential; others simply would rather not take part in the game of life. Such people will not fulfil any 'mini purposes' and therefore will play no part in the fulfilment of the purpose of the human species. Of course, such people will still have their own *individual* purposes which they strive to fulfil.

Why is there such a strong positive correlation between humans fulfilling their potential and humans fulfilling 'mini purposes'? This correlation arises, I suggest, because *the striving of life to survive and human potential are one and the same thing.* If this is so, then as the striving of life to survive and the fulfilling of 'mini purposes' are one and the same thing, this means that when a human fulfils their potential they will automatically be fulfilling 'mini purposes'.

So, why is it the case that the striving of life to survive and human potential are one and the same thing? In order to see why this is so we need to consider in more detail the processes which are involved when life strives to survive. Let us start by considering the first planetary life-forms that arose. Our ongoing assumption has been that these life-forms wanted to stay in existence. In other words, they had a desire to stay in existence which resulted in them striving to stay in existence. This desire to stay in existence would have been passed on to their offspring, who in turn would have passed it on to their offspring. One can see that this desire to stay in existence will still be present in all of the life-forms which exist on the Earth today.

What exactly is this desire? I suggest that at a broad level the desire to stay in existence can be thought of simply as a desire to stay in a good state of feeling – *for the universe* being alive feels good and not being alive doesn't feel good. In other words, *life* is a good state of feeling for the universe to be in; so, when life arises it spreads out from its point of origin and increasingly turns lesser states of feeling into better states of feeling (or turns the absence of feeling into feeling). More specifically this desire can be thought of as *a desire to expand, a desire to propagate, and a desire to explore*. If life is satisfying its desire then one should literally think of life as containing feelings of elation; this part of the universe can be contrasted to the non-living universe which is devoid of these feelings.

How does this desire, which exists within all life-forms, manifest itself within the human species? It will manifest itself as *certain* intuitions/feelings/motivations within an individual human. If a human is acting in accordance with these intuitions/feelings/motivations then they will feel happy/elated/content. If they are not so acting then they are more likely to be unsatisfied/sad/miserable. In short, if a human is acting in accordance with their intuitions/feelings/motivations they will be utilising their talents and abilities; that is to say, they will be fulfilling their potential. Recall that on *Page 54* I made a distinction between *desires,* which pervade the body, and *thought,* which exists at a 'higher level' and is wholly devoid of desire. If one is acting in accordance with one's intuitions/feelings/motivations then one's thought pro-cesses and decisions will be in alignment with the intui-tions/feelings/motivations that pervade one's body. So, it is possible for thinking parts of the universe to contain suppressed desires, whereas desires are insuppressible in non-thinking parts of the universe.

Clearly, the desire to stay in existence will manifest itself as different sets of intuitions/feelings/motivations in different life-forms. Whilst the desire will be the same in say a human and a cat these two life-forms will have different sets of intuitions/feelings/motivations; that is to say, the actions that enable a cat and a human to fulfil their potential are typically different. Indeed, every individual life-form will, to some extent, have a unique set of intuitions/feelings/motivations.

What exactly are intuitions/feelings/motivations? They are states which are internal to an individual which guide them to live their life in a certain way. When one is doing a certain activity, or living one's life in a certain way, one might experience a tingling or a glowing sensation. This is likely to be a sign that one is acting in a good way, one is acting in accordance with one's intuitions/feelings/motivations and is fulfilling one's potential. I often experience such a sensation when I am writing. Whilst, if one is doing something that one really shouldn't be doing, something that one intuitively knows is bad for one, below one, then one is more likely to feel sadness, emptiness and a gnawing sense of unease. If one continuously acts in this way it is likely that one will develop adverse physical health. In the past I have experienced such feelings when doing jobs that weren't utilising my abilities. *As these states are internal only the individual whose states they are can know how they should live their life.*

Let us return to the bigger picture. There are a plethora of 'mini purposes' which need to be fulfilled if the human species is to fulfil its purpose. As we have seen the 'mini purposes' which are available to be fulfilled vary through time (so, one cannot invent the bicycle if the wheel has not been invented). A particular human will be born at a particular stage of cultural evolution and this stage will provide a certain range of possibilities for them to act in accordance with their intuitions/feelings/ motivations. These possibilities are, of course, the 'mini purposes' which can possibly be fulfilled at that time. If a human acts in accordance with their intuitions/ feelings/motivations they will both be fulfilling their potential and fulfilling 'mini purposes'. This is because the intuitions/feelings/motivations of a human *are* the desire of life to survive operating through that human. And it is the desire of life to survive which gives rise to the 'mini purposes' which need to be fulfilled if planetary life is to survive. It is the desire of life to survive – the desire to expand, to propagate, to explore – which is the force which both brought the human species into existence and which gives rise to the trajectory of cultural evolution from hunter-gather to technological society.

So, individual humans do have a purpose, a role to play in the fulfilment of the purpose of the human species. The purpose of individual humans is to fulfil their potential, to live in accordance with their intuitions/feelings/ motivations, to utilise their abilities, to be content and satisfied, and thereby to satisfy the desire of life to survive through the fulfilment of 'mini purposes'. As the majority of individual humans fulfil their potential they each play a part in fulfilling the purpose of the human species.

Chapter 7

Seeing the universe as beautiful

Why do humans often perceive parts of the universe to be beautiful? A human can perceive a stunning sunset, a shimmering ocean, a beautiful rainbow, an enchanting forest, a delicate rose, an exquisite human and the majestic flight of an eagle. Of course, a human can also perceive parts of the universe to be far from beautiful – the hideous sculpture, the grotesque faeces, the offensive language, and the repulsive sight of a pack of vultures devouring its prey; the question is: could it have transpired that the human species evolved to see the entire universe in this way? Could humans inhabit a universe that is wholly devoid of beauty?

It is a well-known phrase that 'beauty is in the eye of the beholder' and this is surely true. If there were no beholders, no eyes, no perceivers of the world, then it would barely make any sense to say that the universe contained parts which are beautiful. Of course, one's perceptions are not limited to seeing with one's eyes; one can hear a beautiful angelic voice and one can also hear a sound which one finds to be highly unpleasant. I am using the phrase *seeing the universe as beautiful* to refer to all of the ways in which one perceives the universe. I suggest that all beauty is in the 'eye' of the beholder; this means that the perceiver-less world is wholly devoid of beauty. On this view it is possible that the human species could have evolved to perceive the entire universe as ugly/repulsive/grotesque. But, clearly this didn't happen; a human can perceive large parts of the universe as beautiful. Why is this?

The answer seems to be obvious. If humans perceived everything as ugly/repulsive/grotesque then it seems fairly safe to assume that their existence would not be a happy one. A life-form which perceived all of its surroundings in this way would have little motivation to procreate and is likely to be miserable, have ill health and a short lifespan. Not only this, one can imagine that if everywhere one went one saw increasing amounts of ugliness/repulsiveness/grotesqueness then it is highly likely that one would not explore the world much; it is likely that one would hide away in a small place; one would probably become a recluse. One can easily imagine such a species becoming extinct fairly soon after it evolved.

Seeing the universe as beautiful seems to be a necessity if a life-form, and a species, is to be evolutionarily successful. Beauty provides the attraction which causes a life-form to act in ways which are good for itself, good for its species, and good for planetary life. I suggest that it is inevitable that a species that has been in existence for a long stretch of time will see the universe as beautiful. Every human will find some other humans to be beautiful, some animals to be beautiful and many other parts of the Earth to be beautiful; in general, they will find many parts of the universe to be beautiful.

In *Chapter Six* I proposed that the desire of planetary life to survive exists in individual humans as particular sets of intuitions/feelings/motivations. I suggested that there are a range of forces in existence which cause the majority of humans to act in accordance with these intuitions/feelings/motivations and thereby simultaneously fulfil their potential and contribute to the purpose of the human species.

In this chapter I am suggesting that seeing the universe as beautiful is another piece of the jigsaw. Seeing the universe as beautiful is another means by which the universe (and, in particular, that part of the universe that is planetary life) ensures that the majority of humans fulfil their potential. The universe evolves the human species in a way which ensures that humans see the universe as beautiful. Beauty is one of the means by which the universe can ensure that the majority of humans act in accordance with their intuitions/feelings/motivations and thereby fulfil their potential; beauty is also the source of many of the intuitions/feelings/motivations that exist in a human.

The beautiful can make a human feel elated, the beautiful can motivate a human, and the beautiful can satisfy a human. When one is frequently surrounded by beauty one is likely to be healthy/content/fulfilled. The beautiful also makes humans want to explore and investigate the world. The existence of the beautiful makes it more likely that humans will fulfil their potential and thereby fulfil some of the 'mini purposes' which need to be fulfilled if the purpose of the human species is to be fulfilled.

In the view I have been forwarding every human has a unique set of intuitions/feelings/motivations endowed to them by the universe. This means that whilst there might be a general agreement amongst humans that certain parts of the universe are beautiful, it will also be the case that some humans will find *particular* things to be beautiful whilst others will not. Some humans will instinctively be drawn towards certain activities and objects, whilst others will not be so drawn. This is what one would expect given that there are a diverse range of 'mini purposes' which need to be fulfilled.

So, what lessons can be learned from the fact that one sees large parts of the universe as beautiful. One can conclude that one sees things as beautiful for a reason – they are good for one. So, you should try to surround yourself with those parts of the universe that you see as beautiful. This will be good for you, it will be good for the human species, and it will be good for planetary life too.

Chapter 8

Respect for all life-forms

Throughout this book I have been claiming that the human species is special; the human species is *not* just another species of animal. There is a good chance that you will agree; indeed, you probably believed this *before* you came across this book. However, I have been claiming that the human species is special because it has a purpose and this is more controversial. According to contemporary conventional wisdom the human species does not have a purpose; on this view the human species is special because the human species has at least one 'special' unique attribute. According to this 'purposeless' view life-forms simply evolve to fit the environmental niches which exist on a planet and there is no overall direction to the evolution of life with the human species at the summit. Furthermore, *irrespective* of the issue of whether the human species has a purpose, and *irrespective* of the issue of whether the universe had a creator, there is a very widespread contemporary belief that the human species is special because the human species has such a 'special' unique attribute.

I have contended that the human species is special but that this widespread contemporary belief is wrong; it is not true that the human species is special because it has a 'special' unique attribute. There is no great chasm in the universe between the human species and everything else; it is simply the case that every species has a bundle of different attributes and that some of the attributes in this bundle are advanced *compared to the degree of advancement in other species* whilst others are not. This means that if the human species is special it must be because it has a purpose. I have proposed that the human species is special because it has a purpose of vital importance – it is the species which planetary life has been striving to attain since it arose. This view of the human species is controversial largely because many humans believe that the evolutionary process is contingent; that is to say, they believe that things could have turned out differently. So, you will have probably heard someone claim that if the asteroid that supposedly caused the dinosaurs to go extinct had just missed the Earth that the dinosaurs (or their evolutionary descendants) could still be the dominant species on the planet. The collision of the asteroid with the Earth was a contingent event; it could have quite easily just missed the Earth with the result that the evolutionary future of the planet would have been very different. I suggest that the evolution of planetary life is a process which is pervaded with such contingencies.

How can the human species be the pinnacle of the evolutionary progression of planetary life if that evolutionary process is pervaded with contingency? Let me explain. If the asteroid that supposedly caused the extinction of the dinosaurs had missed the Earth then the evolutionary progression of planetary life would have been different; the human species might never have evolved. However, I suggest that the desire of life to survive would still have resulted in the evolution of a technological species. The human species is the pinnacle of the evolutionary progression of planetary life; but if things had turned out differently then another species might have occupied this position. There is nothing special about the human species *as such* (as we concluded in *Chapter Two*). The human species is special because it is the part of planetary life that has become technological; it is, therefore, that part of planetary life which has the purpose of saving planetary life.

So, I have claimed that the human species is special; the human species is the peak of the evolutionary summit (as things have turned out on this planet); the human species is the species which planetary life has been striving to bring into existence; the human species is the saviour of planetary life. Now, I wouldn't want all this talk of the human species to give one the wrong impression about the other life-forms on the planet. One should not think of the other life-forms on the planet as non-special, or as our inferiors, or as resources to be used by us.

Is the human species special?

The view of the place of the human species in the evolutionary process that I have outlined in this book entails that it was inevitable that non-human life-forms would come to be used as resources by the human species. Planetary life has been striving for a technological species in order to survive; the human species is that technological species. In order to become technological a species needs to have a certain mentality. It needs to be curious, it needs to explore all of its surroundings (including non-human life-forms) in great detail; and it needs to utilise the resources which are at its disposal. If the human species had not used animals in agriculture, and in other areas, its cultural evolution would have been severely impeded; without such use it is extremely unlikely that the human species would now be as technologically advanced as it is; without such use it is unlikely that the human species would even be a technological species.

From a personal perspective I strongly disapprove of factory farming and vivisection, this is because I value the striving for existence and the individual intuitions/feelings/motivations of all life-forms. But when I consider the broader picture of evolutionary changes that have taken place I am inclined to believe that these things were, perhaps, inevitable stages that needed to be passed through. These things, whatever my personal feelings, seem to be good things for life as a whole. Anything that speeds up progress towards and within a technological society is in the interests of planetary life. Perhaps one just has to accept that certain things, whilst unpleasant in themselves, sometimes serve a bigger purpose.

So, there is a sense in which all parts of the Earth are resources which it is acceptable for the human species to use. It even makes some sense to say that it is in the interests of individual non-human life-forms (even if some or all of them are not aware of the fact) to be used as resources by humans. That is to say, if these life-forms realised that their being used was fulfilling the purpose of saving planetary life, and by being used their distant offspring might have a chance of living, then they might be willing to be used as resources by humans.

Having said all this I urge one *not* to *think of* non-human life-forms as resources to be used by the human species. This way of thinking of non-human animals may have been inevitable in the past, inevitable to get the scientific and industrial revolutions kick-started, but there is no need to think this way today. Today humans are more knowledgeable about their similarity to non-human life-forms. Humans alive today *should think of* non-human life-forms as their companions, and if they are to be used as resources one should take the attitude that these non-human life-forms have willingly sacrificed themselves to serve the purpose of planetary life. I say this because I would like to see the amount of suffering in human society, and within planetary life as a whole, minimised. I am *not* suggesting that humans should never use non-human animals as resources. I *am* suggesting that humans shouldn't think of non-human animals as resources. If a life-form is thought of as an individual with internal motivations and strivings; thought of as a companion part of planetary life; then it will be treated more respectfully and the suffering within planetary life will be reduced. So, one should respect all life-forms.

From the perspective of the view which I have put forward in this book one can see that all life-forms should be cherished. In the universe life is rare; life is precious; life strives its hardest to stay in existence. When an individual life-form dies it is a cause for sadness; when it survives it is a cause for celebration. Of course, it is true that life-forms will consume other life-forms in order to survive; this is inevitable. But when the 'stronger' life-forms consume the 'weaker' life-forms this strengthens the position of planetary life as a whole because 'stronger' life-forms are more likely to survive. This is a good thing, not a bad thing. It is when an individual life-form dies without serving a purpose (that is, when it dies without boosting the survival chances of planetary life) that this is a cause for sadness. I am, of course, here referring to the broad view of what is good or bad for planetary life. From one's personal perspective it can be correct to say that the devouring of one animal by a stronger animal is a bad thing. One might understandably feel this way if one has a close connection to the animal which has just been devoured. For example, one would presumably be very upset if one's pet cat was devoured by a coyote.

The fact that life-forms devour other life-forms, and the specialness of the human species, do not jointly lead to the conclusion that it is right for humans to think of non-human life-forms as resources to be used. As I have said this view might have been inevitable *in the past*. But today it is more appropriate to appreciate the commonality of all life; all of the life-forms on Earth are joined together as different parts of planetary life. When one thinks of things as resources one is more likely to mistreat them, not to consider their perspective, and thereby to bring unnecessary suffering into the world. And it is well known that once a little suffering is brought into existence it spreads, just like when you are kind to someone they are more likely to be kind to someone else. This phenomenon could well be exacerbated by the deeply interconnected nature of the universe; suffering could be radiating out from its point of origin to a multitude of places. If you want to help reduce the amount of suffering that exists on the Earth, and make it a more peaceful place to live, then you should try your hardest to treat all life-forms with care and respect.

Chapter 9

Conclusion

I have presented a particular view of the place of the human species within an evolving universe. This view includes several interlinking elements:

- The human species is special.

- The possession of a unique attribute does not make the human species special.

- The human species has a special purpose – to be the saviour of planetary life.

- Planetary life has been striving to give rise to a technological species ever since it arose.

- There are two forces at work in global culture – the *force to environmental destruction* and the *force to environmental sustainability*. The *force to environmental destruction* is currently overwhelmingly dominant and this is the way things have to be at this moment in time.

- The environmental crisis, and human-induced global warming in particular, is in the interests of planetary life. It is the means by which planetary life ensures that the human species fulfils its purpose.

- If you want to reduce suffering and live on a more peaceful planet you should have respect for all life-forms.

- The best way for one to live is to act in accordance with one's intuitions/feelings/motivations, to do what satisfies one, to utilise one's talents and abilities, to fulfil one's potential.

- Humans see the universe as beautiful because this is a means by which the universe ensures that the majority of humans act in accordance with the intuitions/feelings/motivations that the universe has instilled in them.

- In order to minimise the deleterious effects of human-induced global warming the human species needs to fulfil its purpose as quickly as possible. That is to say, the technology that can regulate the temperature of the planet's atmosphere should be developed and deployed as quickly as possible.

- If the human species fails in its purpose then the future of planetary life is bleak. In the not too distant future only the simplest of life-forms will be able to survive.

If you have turned to this conclusion without reading the rest of the book then you might not be overly convinced by all of these things. However, there is a chance that if you have read the whole book that you might see how all of these parts interlink to form the overall view of the human species that I have presented. You might even agree with this view.

I find the view to be very compelling. It is grounded in a few very plausible premises. Firstly, the Earth and the Sun, like all other parts of the universe, are ageing entities. Secondly, the universe is divided into two parts – life and non-life. Thirdly, life is a good state for the universe to be in. Fourthly, life, and complex life in particular, require certain conditions in order to survive. Fifthly, when life arises it strives to stay in existence by spreading out over the planet it arises on and by regulating the temperature of that planet's atmosphere to keep it favourable for its continued existence. Sixthly, as the Sun's energy keeps on increasing the point will come when, in the absence of a technological species, the ability of planetary life to regulate the temperature of the planet's atmosphere will cease. Seventhly, in order to survive life needs to evolve a technological species. Eighthly, on the Earth the human species is that part of planetary life which is technological. Ninthly, the purpose of the human species is to be the saviour of planetary life through developing and deploying the technology which regulates the temperature of the Earth's atmosphere. Tenthly, the human species is special.

I have claimed that the overwhelming majority of humans have a sense that the human species is special and have contended that the singular source of this sense of specialness is advanced tool-use. However, this sense of human specialness has typically resulted in the belief that humans are the possessors of a unique attribute – a soul, the capacity to think, the ability to feel pain/have emotions, awareness, tool-use, morality, language or culture. I have suggested that neither these attributes nor any other attributes make the human species special. This means that in order to be special the human species needs to be the only part of the universe to have a purpose, or it needs to have a special purpose. I have proposed that the human species is special because it has a special purpose.

Before you came across this book you might have rationalised that the human species is special because it has the ability to create complex technology. Isn't this ability a unique attribute which makes the human species special? I have proposed that if the human species does not have a purpose then complex technology cannot make the human species special. Many species of animals have unique attributes and this would be an arbitrary way of making the human species special. So, the ability to create complex technology does make the human species special *but only because* this ability is required in order to fulfil its purpose as a species – its vital purpose which is to be the saviour of planetary life. It is likely that there are other human attributes, such as advanced morality, which are also required if the human species is to fulfil its purpose. All of the attributes which enable the human species to fulfil its purpose are very valuable attributes.

I have suggested that there are forces in existence – the intuitions/feelings/motivations within humans – that will ensure that the human species fulfils its purpose. In other words, the technology that will regulate the temperature of the Earth's atmosphere will eventually be deployed. If I am right, then in order to reduce human suffering, and the suffering of non-human life-forms, the best strategy is to develop and deploy the technology sooner rather than later. Doing so would deal both with human-induced global warming and with the vastly greater threat from non-human-induced global warming. The intuitions/feelings/motivations which ensure that the human species fulfils its purpose are the forces which underpin the *force to environmental destruction*. I have contended that this force cannot be offset by measures such as the development of renewable technologies, recycling and reductions in resource use per head; the desire of planetary life to survive (which is manifested in the intuitions/feelings/

motivations of humans) is just too strong. Renewable technologies are very valuable but the survival of planetary life requires a very different solution – technological atmospheric temperature regulation.

Of course, it is possible that in the future the human species will not develop and deploy the technology which regulates the temperature of the Earth's atmosphere. This could mean one of two things. It could mean that the human species has failed in its purpose and that planetary life has been decimated. Or, it could mean that the overall view of the human species presented in this book is utterly wrong; there never was such a purpose; this whole scenario being the creation of an overactive imagination.

Nevertheless, one thing *does* seem to be certain. *Even if* the human species does not have a purpose, *even if* individual humans do not have a purpose, *even if* planetary life has not been striving to give rise to a technological species, then *the human species still has the potential to be the saviour of planetary life*. For, in an evolving universe it is only a matter of time before the Earth becomes inhospitable for complex life-forms. One might be under the delusion that it will be billions of years before this time comes. This *is* a delusion.

It *is* true that in a few billion years the Sun will stop ageing and will cease to exist. This means that if planetary life is to survive in the *distant* future then that part of planetary life which is technological needs to transport planetary life-forms to other parts of the universe. However, in this book I have not been focusing on the distant future; my concern has been the *immediate* future.

The fact is that the Sun is getting increasingly hotter. Since life arose on Earth to the present day there has been a 25% increase in solar energy reaching the Earth; in the future this percentage will be *much* higher. Planetary life has been acting so as to offset this increase and thereby keep planetary temperatures favourable for its continued existence. The point that needs to be realised is that, in the absence of a technological species, life cannot continue to regulate the planetary temperature for much longer; its ability is *already* significantly weakening; and human actions have, so far, exacerbated and accelerated this weakening. When will the technology which regulates the temperature of the Earth's atmosphere be required? Unfortunately, the exact timing cannot be known; this is because we do not understand the ageing processes of the Earth and the Sun well enough, *and* we do not understand the way in which the Earth's biosphere works well enough to know if a 'tipping point' is going to be passed or a 'runaway greenhouse effect' is going to occur. However, it is clear that we are not dealing with a timeframe of millions of years; the technology will be required in thousands, if not hundreds, of years.

I like to hope that if the human species is not technologically regulating the temperature of the Earth's atmosphere by the end of the millennium (the year 3000) that the Earth might still be home to wonderfully complex life-forms; but I could be wrong.

Introduction to the appendices

It was in early 2005 that I came up with the overall view of the place of the human species within an evolving universe that I have presented in this book. Later that year I was enthralled when I came across the views of Friedrich Hölderlin; here was someone who seemed to have a very similar view. Hölderlin formulated his view of the human species and its place in universal evolution at the end of the eighteenth century, long before the onset of the environmental crisis. In Hölderlin's era there were definitely no concerns about human-induced global warming. Contrarily, the environmental crisis was *central* to my thinking about the place of humanity in universal evolution. After studying Holderlin's works, which I enjoyed greatly, I was sure that if he were still alive he would agree with my view of the environmental crisis. So, I wrote a paper on this subject and this paper was published in the journal *Cosmos and History: The Journal of Natural and Social Philosophy* in 2007 (Vol. 3, No. 1).

This paper is to be found on the following pages as Appendix A. When I wrote the paper I seem to recall that I found myself in agreement with all of Holderlin's views. However, as you will see, Holderlin seemed in the end to conclude that the human species has no freedom; that is to say, it is not possible that a human could have ever acted any differently. Today I am inclined to think that it is possible to have the view of the human species that I have outlined in this book whilst still believing that humans have the freedom to do otherwise. That is to say, one can believe that it is the universe which provides humans with their intuitions/feelings/motivations but that one has the freedom either to act or not act in accordance with them.

In 2008 I won an international prize essay contest which was run by the Spinoza-Gesellschaft. There was a set title for the contest: *How much of man is natural?* In the title the word 'man' is obviously being used to refer to the human species as a whole. In this essay I address similar themes to those addressed in *Chapter 2*. In the essay I contend that all parts of the universe are unique and natural; whereas in *Chapter 2* my focus is solely on the life-forms which exist on the planet Earth. In the essay I also explore the concept of naturalness and consider why it is that humans often do not consider themselves to be wholly natural. A slightly modified version of this essay is to be found in the following pages as Appendix B.

If you have enjoyed reading the book so far then there is a good chance that you will also enjoy these two pieces of work. They nicely complement the view of the human species as an important part of an evolving universe that I have presented in the main part of this book. Indeed, one will probably conclude that there is a strong link between what I have been referring to as *the human species fulfilling its purpose,* the epoch Hölderlin refers to as *the absence of the gods,* and *humans not considering themselves to be natural.* In short, it is perhaps due to the absence of the gods that humans do not consider themselves to be natural, and, in turn, perhaps this view of themselves is required in order for humans to fulfil their purpose.

Appendix A

Human nature, cosmic evolution and modernity in Hölderlin

Abstract: The German Romantic Friedrich Hölderlin developed a unique perspective on the relationship between humankind and the rest of nature. He believed that humanity has a positive role to play in cosmic evolution, and that modernity is the crucial stage in fulfilling this role. In this paper I will be arguing for a reinterpretation of his ideas regarding the position of humankind in cosmic evolution, and for an application of these ideas to the 'environmental crisis' of modernity. This reinterpretation is significant because it entails an inversion of the conventional notion of causality in the 'environmental crisis'; instead of humans 'harming' nature, in the reinterpretation it is nature that causes human suffering.

Keywords: Human Nature, Organismic Evolution, Fate, Culture, Technology, Modernity, Environmental Crisis, Human Suffering, Anthropocentricism.

Friedrich Hölderlin, one of the German Romantics, developed a distinctive viewpoint on the relationship between humankind and the rest of nature. His ideas are of particular interest because he yearned for an end to human suffering, but was also firmly convinced that humankind was inevitably destined to be separated from nature, and thereby destined to endure suffering. Hölderlin envisioned a positive role for humanity in cosmic evolution, a role which has significant implications for both human nature and cultural evolution. In this paper I will be outlining Hölderlin's ideas, and arguing for an application of them to the 'environmental crisis' of modernity. Hölderlin's conception of the human-nature relationship as part of an unfolding process of cosmological change seems to be of great relevance today, an age that is characterized by belief in the meaninglessness of human existence, and by concern about the way that we have altered the pre-human conditions of the Earth. Hölderlin's views provide a unique perspective on modernity that is worthy of serious consideration.

I start by outlining Hölderlin's views on the role of humankind in universal evolution. I then review the secondary literature on Hölderlin that relates to these ideas. I proceed to argue that Hölderlin's philosophy is applicable to, and gives a unique perspective on, the 'environmental crisis' of modernity. I argue that the existing secondary literature on Hölderlin has not recognized this, and that a reinterpretation of the role of humanity in Hölderlin's philosophy of cosmic evolution is therefore required. My central claim is that for Hölderlin, modernity and the related notion of the contemporary 'environmental crisis' is a necessary stage of cosmic evolution, and thus that it is far from a 'crisis'. Rather it is a necessary stage of disharmony that will inevitably be followed by a re-conquered harmony. I will argue that for Hölderlin this disharmony is characterized by the environmental changes that are resultant from the development of technology.

1. *Hölderlin's philosophy of human nature, cosmic evolution and modernity*

The starting point of Hölderlin's philosophy is that there must be a basic unknowable reality which precedes self-consciousness wherein subjects and objects are not in existence but are both part of a 'blessed unity of being'. He describes this unity as, "Where subject and object simply are, and not just partially, united...only there and nowhere else can there be talk of being."[1] He argues that the 'blessed unity of being' (which he also refers to as 'nature') is responsible for the coming into existence of humanity through using its power to initiate a division of itself into subjects and objects. This division of being causes the emergence of judgement. Hölderlin states that, "'I am I' is

---

[1] Friedrich Hölderlin, 'Being Judgement Possibility', in J. M. Bernstein (ed.), *Classic and Romantic German Aesthetics,* Cambridge, Cambridge University Press, 2003, p. 191.

the most fitting example of this concept of judgement...[as] it sets itself in opposition to the *not-I,* not in opposition to *itself.*"[2]

The division means that human beings are not capable of actions that are independent of nature; Hölderlin states that, "all the streams of human activity have their source in nature."[3] It is revealing to compare this claim with the words of Hölderlin's character Hyperion, "What is man? – so I might begin; how does it happen that the world contains such a thing, which ferments like a chaos or moulders like a rotten tree, and never grows to ripeness? How can Nature tolerate this sour grape among her sweet clusters?"[4] For Hölderlin, man is the 'violent' being, whose coming into existence in opposition to the rest of nature was *initiated* by nature.

Hölderlin sees this opposition between man and the rest of nature as culminating in modernity – an era that he claims is characterised by the absence of the gods. In *Brot und Wein* Hölderlin writes, "Though the gods are living, Over our heads they live, up in a different world...Little they seem to care whether we live or do not."[5] A key question for Hölderlin is how we deal with this separation.

---

[2] Ibid., p. 192.

[3] Alison Stone, 'Irigaray and Hölderlin on the Relation Between Nature and Culture', in *Continental Philosophy Review,* vol. 36, no. 4, 2003, p. 423.

[4] Friedrich Hölderlin , 'Hyperion', in Eric L. Santner (ed.), *Hyperion and Selected Poems,* New York, Continuum, 1990, p. 35.

[5] Ibid., p. 185.

He envisions two possibilities – the 'Greek' response which is to dissolve the self and die, and the 'Hesperian' response of a living death.

Hölderlin came to view the 'Greek' response as hubristic, it being based on an anthropocentric desire to oppose the division initiated by nature. He thus sees the 'Hesperian' response of living and carrying out actions that are dependent on nature for their origination as the appropriate non-hubristic response to our separation. Hölderlin's position is that as nature created the separation, *only* nature can bring the separation to an end. He sees this process of separation and reconnection as part of a broader cosmic picture wherein nature is an unfolding organism rather than a huge mechanism. This organismic view enables him to envision teleological processes in nature which give rise to his claim that there will be, "eternal progress of nature towards perfection."[6]

2. Interpretations of Hölderlin and his concept of fate

In this section I set out my view of Hölderlin's conception of fate – that all human actions are part of the evolution of nature towards perfection. I do this by reviewing the existing scholarly literature on Hölderlin and showing that whilst these interpretations all recognise parts of Hölderlin's conception of fate that they do not capture the whole

---

[6] Ronald Peacock, *Hölderlin,* London, Methuen & Co. Ltd, 1938, p. 36.

of it. I start with interpretations of human nature, move on to cosmic processes, and finally consider the role of modernity within these processes.

At the level of the human there is a general consensus in the literature that Hölderlin's position is that humans are endowed by nature with qualities that shape human nature, and that this inevitably shapes human interactions with the rest of nature. There are various names in the literature for the qualities which are endowed to humans. Dennis J. Schmidt refers to the qualities present in humans as their 'formative drive.' He claims that, "Hölderlin suggests that human nature and practices are to be understood by reference to a formative drive which expresses itself as a constant need for 'art'."[7] In a similar vein, Thomas Pfau argues for an 'intellectual intuition.' He states that, "Hölderlin recasts the convergence of "freedom and necessity" as the most primordial synthesis of intellect and intuition itself, a synthesis which takes place within the subject itself. He thus approaches what Kant had repeatedly ruled out as an "intellectual intuition"."[8]

In agreement with Schmidt and Pfau, Franz Gabriel Nauen argues that for Hölderlin, "all men do in fact have the same basic character...all human activity can be

---

[7] Dennis J. Schmidt, *On Germans and Other Greeks,* Indiana University Press, 2001, p. 139.
[8] Thomas Pfau, *Friedrich Hölderlin: Essays and Letters on Theory,* New York, SUNY Press, 1988, p. 15.

derived from the same *elemental drive* in human nature."[9] The 'formative drive' / 'intellectual intuition' / 'elemental drive' identified in the literature explains why man can be seen as the 'violent' being. Human nature is to engage in 'art', to utilize the resources of nature so that culture can be generated and sustained. This generation of human culture actually benefits nature as a whole, but it requires large-scale modification of parts of non-human nature. The destiny of man is thus a disruptive one. It is clear that it is also an undesirable one. Nauen states that for Hölderlin, "Even war and economic enterprise serve to fulfil the destiny of man, which is to "multiply, propel, distinguish and mix together the life of Nature"."[10]

So Hölderlin sees human nature, economic production and even war as parts of a broader cosmic evolutionary process; the universe *as a whole* is seen as evolving to perfection. There will inevitably be aspects of this evolution that from a narrow perspective could be viewed as 'less than perfect'. These negative aspects of the evolutionary process – from war, to the presence of evil in its entirety – have to be seen as inescapable parts of the whole process.

The key point is that for Hölderlin the cosmic evolutionary process *ends* in perfection. Thus, Ronald Peacock

---

[9] Franz Gabriel Nauen, *Revolution, Idealism and Human Freedom: Schelling, Hölderlin and Hegel and the Crisis of Early German Idealism,* Indiana University Press, 2001, p. 139.
[10] Ibid.

argues that, "the division produced by conflict is followed by a re-conquered harmony."[11] Similarly, Anselm Haverkamp argues that an interpretation of the poems *Andenken* and *Mnemosyne* is the expression, 'where danger threatens, salvation also grows.'[12] Whilst, Martin Heidegger translates the opening lines of *Patmos* as, "But where danger is, grows the saving power also."[13] Hölderlin's view is clearly that from a narrow and short-term perspective danger and conflict are often the norm, but that these things actually play a part in bringing about a greater harmony in the future. In the long-term they are all part of the evolution of the whole universe to perfection.

Cosmic evolution is thus one long process of disharmonies and inevitably following harmonies. Peacock argues that Hölderlin's vision is of a, "harmonised process of life which comprises within itself the rhythmic movement from chaos to form and back again, and an emotional experience of this which in the sphere of nature knows only the one rapture, but in the human sphere suffering and joy."[14] It is revealing that this interpretation sees 'violent' humans as suffering, whilst nature is purely

---

[11] Peacock, *Hölderlin,* p. 22.

[12] Anselm Haverkamp, *Leaves of Mourning: Hölderlin's Late Work,* New York, SUNY Press, 1996, p. 48.

[13] Martin Heidegger, 'The Question Concerning Technology', in R.C. Scharff and V. Dusek (eds.), *Philosophy of Technology: The Technological Condition – An Anthology,* Oxford, Blackwell Publishing, 2003, p. 261.

[14] Peacock, *Hölderlin,* p. 22.

rapturous. This clearly sheds light on the question posed by Hölderlin's character Hyperion: "How can Nature tolerate this sour grape among her sweet clusters?"[15] The answer seems to be that human 'violence' *enables* nature to be rapturous. As part of this rapture humans experience suffering.

Why should suffering be a uniquely human experience? To explain this Peacock cites part of a letter from Hölderlin to his brother, "Why can they [humans] not live contented like the beasts of the field? he asks: and replies that this would be as unnatural in man, as in animals the tricks, or arts, man trains them to perform. Thus he establishes that the arts of man are natural to man. Culture, then, derives from nature; and the impulse to it is the characteristic which distinguishes man from the rest of creation."[16]

The human impulse to culture has culminated in the era of modernity. Hölderlin sees this period as one of great significance as he sees it as a historical epoch that is characterised by the *absence of the gods*. To be consistent with his views on harmonised evolution to perfection there must be a reason for this absence. Indeed, Peacock argues that Hölderlin thinks that, "a godless age is part of a divine mystery, it is as necessary as day, ordained by a higher

---

[15] Hölderlin, 'Hyperion', p. 35.

[16] Peacock, *Hölderlin,* p. 36.

power."[17] Furthermore, Heidegger claims that the gods are still present, despite their absence: "man who, even with his most exulted thought could hardly penetrate to their Being, even though, with the same grandeur as at all time, they were somehow there."[18]

The absence of the gods in modernity is deeply related to the contemporary danger that exists in modernity. It should be remembered that this danger cannot be a cause for concern for Hölderlin – as all dangers are inevitably followed by regained harmonies. Nevertheless, Heidegger attempts to identify the exact danger that Hölderlin believed is present in modernity. Heidegger claims that, "the essence of technology, enframing, is the extreme danger."[19] It must follow that for Heidegger, "precisely the essence of technology must harbor in itself the growth of the saving power."[20] He sees this as occurring when the essential unfolding of technology gives rise to the possibility of opening up a "free relation" with technology which is inclusive of non-instrumental possibilities.[21]

---

[17] Ibid., p. 92.

[18] Martin Heidegger, *Existence and Being,* London, Vision Press Ltd., 1956, p.190.

[19] Heidegger, 'The Question Concerning Technology', p. 261.

[20] Ibid.

[21] R.C. Scharff and V. Dusek, 'Introduction to Heidegger on Technology', in R.C. Scharff and V. Dusek (eds.), *Philosophy of Technology: The Technological Condition – An Anthology,* Oxford, Blackwell Publishing, 2003, p. 248.

In an interpretation of the 1802 hymn *Friedensfeier,* Richard Unger draws out Hölderlin's views on the absence of the gods in modernity.[22] In *Friedensfeier* the entire span of Western civilization is characterised as a thunderstorm which is ruled by a "law of destiny" which ensures that a certain amount of "work" is achieved. Unger argues that it is clear that this "work", "is the product of the storm itself and that it designates the harmonious totality of earthly existence during the coming era."[23] The end of the "storm" of modernity enables the arrival of a mysterious "prince" who makes it possible that, "men can now for the first time hear the "work" that has been long in preparation "from morning until evening"."[24]

Following the inevitable successful accomplishment of the "work" of Western civilization, the great Spirit will disclose a Time-Image which will, "be a comprehensive depiction of the historical process and its triumphant result."[25] Unger argues that, "the Image shows that there is an alliance between the Spirit of history and the elemental divine presences of nature – for the natural elements with which man has always worked have played integral and essential parts in man's history."[26] The triumphant result

---

[22] Richard Unger, *Friedrich Hölderlin,* Boston, Twayne Publishers, 1984, pp. 100-105.

[23] Ibid., p. 102.

[24] Ibid., p. 101.

[25] Ibid., p. 104.

[26] Ibid., p. 105.

of the actions of humankind in modernity is clearly an example of a re-conquered harmony that follows division.

In Unger's interpretation of *Friedensfeier* we have a picture of modernity in which humans are carrying out "work" under a "law of destiny". The crucial factor is that humanity is ignorant that it is working under a "law of destiny" in modernity, until modernity has ended. It is then that through the Time-Image the great Spirit reveals the successful outcome of modernity, and the *nature and value* of the accomplished "work". This is a prime example of a short-term and narrow perspective entailing the perception of a lack of destiny and of needless suffering, whilst in the longer-term the same events are seen to be an inevitable part of a broader positive outcome – the evolution of the universe to perfection.

This difference of perspectives can explain an apparent contradiction in the literature between Unger's interpretation of *Friedensfeier,* and Schmidt's analysis of Hölderlin's 1801 letter to Bohlendorff. This letter was written only one year before *Friedensfeier* and Schmidt claims that in it Hölderlin's position is, "that the peculiar flow of modernity is the lack of destiny."[27] The apparently contradictory views of Unger and Schmidt can be reconciled through recalling Peacock's interpretation that, "a godless age is part of a divine mystery, it is as necessary as

---

[27] Schmidt, *On Germans and Other Greeks,* p. 137.

day, ordained by a higher power,"[28] and comparing it to Unger's claim that men are blind to the point of the "work" that they have been carrying out until the "storm" of Western civilization has passed.

The comparison reveals that the "law of destiny" applies to the activities of *humanity as a collective* in Western history, activities that are ordained by a higher power for a specific purpose. In contrast, the "lack of destiny" applies to *individual human beings*. This difference arises because individual humans are unaware that their actions are part of an inevitably unfolding cosmic plan, it is only the fruition of the plan than enables realization. Instead, humans believe that they have free will and live in a meaningless age. Therefore, modernity can at one and the same time be characterized as both a period governed by a "law of destiny" and a period constituted by a "lack of destiny". The difference is purely one of perspective.

This conception of modernity as simultaneously being a period of a "lack of destiny" and a "law of destiny" raises the issue of anthropocentricism. If human attitudes and actions towards nature are in the interests of nature, then it seems that there is no such thing as a *truly* anthropocentric attitude. The appropriate attitude that humans should take to the objective side of nature, given Hölderlin's philosophy, has been addressed by Alison Stone. She

---

[28] Peacock, *Hölderlin,* p. 92.

argues that because, "according to Hölderlin's thinking, we have become separated from nature by *its* power alone, so it is not within *our* power to undo separation."[29] Therefore, "the appropriately modest response is to endure separation – to wait, patiently, until nature may change its mode of being."[30] This means that a truly non-anthropocentric environmental view of the rest of nature requires, "the *acceptance* of disenchantment, of separation, of meaninglessness."[31]

This view is concordant with the "lack of destiny" perspective. However, when the "law of destiny" is taken into account, then the hidden meaning is revealed. Furthermore, the whole notion of the attitudes of individual humans then becomes irrelevant. It seems that there cannot be such a thing as a *truly* anthropocentric attitude, because all attitudes originate from nature, and they all lead to actions which fulfil the "law of destiny". It may seem that our attitudes to nature are of importance, but this is because we believe in a "lack of destiny", and are inevitably blind to the bigger picture of the "law of destiny". Whatever our attitudes as individuals, our

---

[29] Stone, 'Irigaray and Hölderlin on the Relation Between Nature and Culture', p. 424.

[30] Ibid.

[31] Alison Stone, *Nature in Continental Philosophy – Week 4, Section V, Friedrich Hölderlin,* [online], http://www.lancaster.ac.uk/depts/philosophy/awaymave/408new/wk4.htm, [accessed 25 October 2005].

relationship with the rest of nature as a collective would be 'for the best'.

3. A reinterpretation of the human in cosmic evolution

The interpretations of Hölderlin that I have reviewed all give an accurate representation of Hölderlin's views. However, they are all partial views. They all miss the 'big picture' of what Hölderlin's views imply about what it means to be a human in the context of cosmic evolution, and the consequent implications for the perspective from which we should view modernity and the 'environmental crisis'. In an attempt to fully grasp these implications I am going to defend the thesis that: *Hölderlin's philosophy leads to the conclusion that the 'environmental crisis' is a necessary stage in the purposeful evolution of nature towards perfection.* This is an interesting thesis because, if accepted, it would supplant the conception of the mean-inglessness of human existence with a conception of positive cosmic purpose.

The argument I will be making centers on three key aspects of Hölderlin's philosophy. Firstly, that he believes that nature is purposefully evolving towards perfection. Secondly, that he believes that the achievement of this perfection requires human actions. Thirdly, that he believes that human actions are determined by nature. Acceptance of these three claims leads to the conclusion that human actions are determined by nature as a necessary stage in the purposeful evolution of nature towards perfection. As the 'environmental crisis' of modernity is purely resultant from human actions, a second conclusion inevitably follows. This is that the 'environmental crisis' itself is determined by nature as a necessary stage in the purposeful evolution of nature towards perfection.

I will now present evidence to support the three key claims. The first claim is that Hölderlin's belief is that *nature is purposefully evolving towards perfection.* The universe can either be viewed as a giant mechanism or as an unfolding organism; Hölderlin clearly held the latter view. This conception of the universe explains his belief that nature unfolds in a way that serves its own purposes; that disharmonies are followed by regained harmonies. This is why Peacock claims that Hölderlin believed in, "the eternal progress of nature towards perfection,"[32] and, "the emergence of perfection in the course of natural development."[33]

This firm belief clashed with Hölderlin's personal yearning for immediate perfection in life. His immense desire to see a morally just world was completely at odds with his philosophical belief that the perfection he sought could only be achieved in the course of natural development. The movement to perfection envisioned by

---

[32] Peacock, *Hölderlin,* p. 36.
[33] Ibid., p. 105.

Hölderlin is thus a fatalistic one, an inevitable evolutionary progression towards perfection. Peacock captures this with his claim that for Hölderlin there is an, "acute sense of 'Fate', of inevitability, expressed again and again in his work. Fate is revealed in the process of history... it is inherent in the passage of form to chaos, and of disintegration to a new harmony."[34]

This first claim is the most straightforward of the three. The second claim is that *Hölderlin believes that the achievement of perfection requires human actions.* The starting point in defending this claim is Hölderlin's central belief that nature *used its power* to divide itself and thereby create humankind. This division means that the split was part of the evolutionary process rather than a random occurrence. We can ask ourselves why this may have been a necessary occurrence. An initial answer seems to be Nauen's claim that, "Even war and economic enterprise serve to fulfil the destiny of man, which is to "multiply, propel, distinguish and mix together the life of Nature"."[35]

In *The Perspective from which we Have to look at Antiquity* Hölderlin asserts that, "antiquity appears altogether opposed to our primordeal drive which is bent on forming the unformed, to perfect the primordial-

---

[34] Ibid., p. 93.

[35] Nauen, *Revolution, Idealism and Human Freedom: Schelling, Hölderlin and Hegel and the Crisis of Early German Idealism,* p. 139.

natural so that man, who is born for art, will naturally take to what is raw, uneducated, childlike rather than to a formed material where there has already been pre-formed [what] he wishes to form."[36] In a letter to his brother he also asserts that, "the impulse to art and culture...is really a service that men render nature."[37]

The source of Hölderlin's primordeal drive to art is nature, because it is nature that created us and endowed us with our capabilities. This is clear from Peacock's interpretation that, "Man cannot be master of nature; his arts, *necessary though they may be in the scheme of things,* cannot produce the substance which they mould and transform; they can only develop the creative force, which in itself is eternal and not their work."[38]

Hölderlin's primordeal drive to art in humans has inevitably led to the epoch of modernity. Human actions in this epoch appear to be central to the achievement of perfection. Hölderlin claims that modernity is an epoch that, "is as necessary as day, ordained by a higher power."[39] Furthermore, humans have been involved in "work" in modernity that is clearly constitutive of the importance of the epoch. This is clear from Unger's

---

[36] Friedrich Hölderlin, 'The Perspective from which We Have to Look at Antiquity', in Thomas Pfau (ed.), *Friedrich Hölderlin: Essays and Letters on Theory,* New York, SUNY Press, 1988, p. 39.

[37] Peacock, *Hölderlin,* p. 37.

[38] Ibid.

[39] Ibid., p. 92.

interpretation of *Friedensfeier* in which the "law of destiny" ensures that a certain amount of human "work" is done. The crucial factor is that humanity is ignorant that it is working under a "law of destiny" in modernity, until modernity has ended. It is then that through the Time-Image the great Spirit reveals the successful outcome of modernity, and the nature and value of the accomplished "work".

There is no doubt that in Hölderlin's view human actions and their resultant "work" in modernity are part of purposeful evolution to perfection. What is interesting is the exact nature of the "work". There is an obvious connection between the "work" of modernity (*Friedensfeier*) and the "danger" we face in modernity (*Patmos*). Heidegger's interpretation of *Patmos* that, "the essence of technology, enframing, is the extreme danger,"[40] makes it clear that the "work" of modernity is the development of technology. In fact, technological development in modernity seems to be the culmination of Hölderlin's primordeal drive to art. Furthermore, it is very hard to think of any other distinctive aspects of modernity that are resultant from human actions, present an extreme danger, and have cosmic significance. Therefore, for Hölderlin, the achievement of perfection seems to require the human development of technology.

---

40 Heidegger, 'The Question Concerning Technology', p. 261.

It is interesting that Heidegger sees the danger we face from the "work" of modernity as the essence of technology rather than actual technology. Andrew Feenberg has criticised Heidegger for this abstract concentration on essences rather than the actual technology itself.[41] A "Feenberg interpretation" of *Patmos* seems to be more in accordance with Hölderlin's views than the "Heidegger interpretation", as Hölderlin's philosophy is grounded in actualities rather than essences. Hölderlin sees a positive role for actual technology in cosmic evolution; this means that *actual technology* has a cosmic purpose. Therefore, it seems that both the danger we face, and the saviour, must be the *actual* technology developed by human actions.

The importance of the human split from the rest of nature can also be seen in the words of Hölderlin's character *Hyperion:* "How should I escape from the union that binds all things together? We part only to be more intimately one, more divinely at peace with all, with each other. We die that we may live."[42] Human actions are thus depicted as a 'living death' that is necessary for the life (and continued movement to perfection) of nature as a whole. This explains Peacock's interpretation that, "the

---

[41] Andrew Feenberg, 'Critical Evaluation of Heidegger and Borgmann', in R.C. Scharff and V. Dusek (eds.), *Philosophy of Technology: The Technological Condition – An Anthology,* Oxford, Blackwell Publishing, 2003, pp. 327-337.
[42] Hölderlin, 'Hyperion', p. 123.

sphere of nature knows only the one rapture, but in the human sphere [there is] suffering and joy."[43]

The third claim is that *Hölderlin believes that human actions are determined by nature.* There are many passages in Hölderlin's novel *Hyperion* that attribute the responsibilities for human actions to a power or god: "There is a god in us who guides destiny as if it were a river of water, and all things are his element."[44]....."oh forgive me, when I am compelled! I do not choose; I do not reflect. There is a power in me, and I know not if it is myself that drives me to this step."[45]....."I once saw a child put out its hand to catch the moonlight; but the light went calmly on its way. So do we stand trying to hold back everchanging Fate. Oh, that it were possible but to watch it as peacefully and meditatively as we do the circling stars."[46]....."Man can change nothing and the light of life comes and departs as it will."[47]....."We speak of our hearts, of our plans, as if they were ours; yet there is a power outside of us that tosses us here and there as it pleases until it lays us in the grave, and of which we know not where it comes nor where it is bound."[48]

---

[43] Peacock, *Hölderlin,* p. 22.
[44] Hölderlin, 'Hyperion', p. 11.
[45] Ibid., p. 79.
[46] Ibid., p. 22.
[47] Ibid., p. 127.
[48] Ibid., p. 29.

Hölderlin's belief in the lack of human free will is perhaps clearest in his claim in a letter to his mother regarding the views of Spinoza that, "one *must* arrive at his ideas if one wants to explain everything."[49] Spinoza's ideas can be summed up as, "Nature in all its aspects is governed by necessary laws, and human being no less than the rest of nature is determined in all its actions and passions, contrary to those who conceive of it as 'a dominion within a dominion'."[50]

In order to make abundantly clear Spinoza's - and thus Hölderlin's – views on a lack of human free will here are two quotes from Spinoza: "I say that thing is free which exists and acts solely from the necessity of its own nature...I do not place Freedom in free decision, but in free necessity."[51] And, "a stone receives from an external cause, which impels it, a certain quantity of motion, with which it will afterwards necessarily continue to move...Next, conceive, if you please, that the stone while it continues in motion thinks, and knows that it is striving as much as possible to continue in motion. Surely this stone, inasmuch as it is conscious only of its own effort, and is far

---

[49] Friedrich Hölderlin, 'No.41: To his Mother', in Thomas Pfau (ed.), *Friedrich Hölderlin: Essays and Letters on Theory,* New York, SUNY Press, 1988, p. 120.

[50] Moira Gatens, *Imaginary Bodies: Ethics, Power and Corporeality,* London, Routledge, 1996, p. 111.

[51] Benedict de Spinoza, 'LVIII: To Schuller', trans. A. Wolf (ed.), *The Correspondence of Spinoza,* 2nd ed., London, Frank Cass & Co. Ltd., 1966, pp. 294-5.

from indifferent, will believe that it is completely free, and that it continues in motion for no other reason than because it wants to. And such is the human freedom which all men boast that they possess, and which consists solely in this, that men are conscious of their desire, and ignorant of the causes by which they are determined."[52]

Furthermore, in an interpretation of Hölderlin's *Stutgard*, Peacock argues that, "the laws of growth govern the culture as well as the lives of men...the one process comprehends all things and the one rhythm manifests itself again and again...in the progress of history; in the spiritual life of individuals."[53] In this vision not only human nature, but also the evolution of culture, is seen as an inevitable historical progression. Peacock's interpretation of Hölderlin is that, "man's spirit is but part of the One Spirit,"[54] which Hölderlin insists is involved in a "movement...through successive historical generations."[55] The spirit of man is thus governed by the larger Spirit of nature. This is the sense in which, "all the streams of human activity have their source in nature."[56]

The nature of the relationship between man's spirit and the Spirit of nature is made clear in the following

---

[52] Ibid., p. 295.
[53] Peacock, *Hölderlin,* p. 25.
[54] Ibid., p. 90.
[55] Ibid., p. 114.
[56] Stone, 'Irigaray and Hölderlin on the Relation Between Nature and Culture', p. 423.

quote from Hölderlin's character Diotima: "a *unique destiny* bore you away to solitude of spirit as waters are borne to mountain peaks."[57] This concept of individual humans having a unique destiny was the view of Johann Herder, who was one of Hölderlin's inspirations. Herder saw nature as a great current of sympathy running through all things which manifested itself in unique inner impulses within different individuals. This means that every human has a unique calling – an original path which they ought to tread. As Herder states, "Each human being has his own measure, as it were an accord peculiar to him of all his feelings to each other."[58] Clearly, for both Herder and Hölderlin, human actions at any one time are determined in accordance with the movements of the One Spirit of nature.

I have presented evidence for the claims that for Hölderlin: *nature is purposefully evolving towards perfection, the achievement of this perfection requires human actions, and human actions are determined by nature.* Acceptance of these three claims leads to the conclusion that human actions are determined by nature as a necessary stage in the purposeful evolution of nature towards perfection. I now briefly argue that the 'environ-

---

[57] Hölderlin, 'Hyperion', p. 122.
[58] Charles Taylor, *Sources of the Self: The Making of the Modern Identity,* Massachusetts, Harvard University Press, 1994, p. 375.

mental crisis' of modernity is purely resultant from human actions.

The definition of an environmental problem is: "any change of state in the physical environment which is brought about by human interference with the physical environment, and has effects which society deems unacceptable in the light of its shared norms."[59] This definition encapsulates a sliding scale of environmental problems from those that are local and temporary on the one hand, to those that are global and long-lasting on the other. The 'environmental crisis' as a concept has arisen because of the emergence in the last 100 years of an increasing number of environmental problems that are towards the global and long-lasting end of the scale. The 'environmental crisis' is thus purely resultant from the *human actions* which have created environmental problems that are characterised by their global reach and long-lasting nature.

This means that the above conclusion, that human actions are determined by nature as a necessary stage in the purposeful evolution of nature towards perfection, needs amending. As the 'environmental crisis' is purely resultant from human actions, it too must be part of this purposeful evolution. Therefore, the new conclusion that

---

[59] Peter B. Sloep and Maris C.E. van Dam-Mieras, 'Science on Environmental Problems', in P. Glasbergen and A. Blowers (eds.) *Environmental Policy in an International Context: Perspectives,* Oxford, Butterworth-Heinmann, 2003, p. 42.

inevitably follows is: *the 'environmental crisis' is determined by nature as a necessary stage in the purposeful evolution of nature towards perfection.*

4. Objections to the reinterpretation

It could be objected that there are many references to human freedom in Hölderlin's work that would seem to cast doubt on the third claim. This is particularly noticeable in his novel Hyperion. For example, Hyperion states that, "without freedom all is dead."[60] However, this objection is easily answered because these references all appear in Hölderlin's early work, and even then they are more than counterbalanced by the opposing fatalistic views that I have outlined. In his early period Hölderlin was struggling to come to terms with the conflict between his keen moral aspirations for social change on the one hand, and his belief in perfection only arising through natural development on the other. In his later work, as is clear in his endorsement of the 'Hesperian' response to our condition, he firmly accepts the powers of natural development and the determination of human actions by nature. He realizes the futility of pursuing his idealistic moral aspirations because he accepts the illusory nature of human free will.

---

[60] Hölderlin, 'Hyperion', p. 117.

A further objection could be made that this reinterpretation is pointless because Darwin's theory of evolution, which emerged shortly after Hölderlin's time, gives a view of evolutionary processes that is incompatible with Hölderlin's view that there was a 'blessed unity of being' prior to the arrival of humans. We now know that the emergence of the human species – and its primordeal drive to art – was preceded by four billion years of evolution of life on Earth. It can thus be argued that there was not a 'blessed unity of being' prior to the evolution of humankind.

This is exemplified by the claim of Hans Jonas that the subject-object divide opened up four billion years ago, when, "living substance, by some original act of segregation, has taken itself out of the general integration of things in the physical context, set itself over against the world, and introduced the tension of "to be or not to be" into the neutral assuredness of existence."[61] This certainly does not appear to be a pre-human 'blessed unity of being'. However, it is interesting that Jonas also sees humans as, "a 'coming to itself' of original substance."[62]

It is clear that this Darwinian based objection does not invalidate the views of Hölderlin, or the reinterpretation of them presented in this paper. In fact, not only does

---

[61] Hans Jonas, *The Phenomenon of Life: Toward a Philosophical Biology*, Illinois, Northwestern University Press, 2001, p. 4.
[62] Ibid., p. xv.

evolutionary theory perfectly complement Hölderlin's philosophy, his philosophy *needs* it. The idea that nature could use its power to instantaneously create a being as complex as a human out of the 'blessed unity of being' is hardly defensible. In the light of our knowledge today we can simply reinterpret Hölderlin as claiming that nature used its power four billion years ago to divide the 'blessed unity of being' and create a subject/object divide. As he sees nature as an unfolding and evolving organism, the divide would give rise to human subjects after a sufficient period of time. This, ""coming to itself" of original substance", as Jonas describes it, has in actuality taken approximately four billion years.

5. Conclusion

I have argued that the existing secondary literature has not grasped the full implications of Hölderlin's thought for what it means to be a human in modernity. By drawing together Hölderlin's ideas I have sought to understand his notion of the purpose of human actions, and what this purpose means for the 'environmental crisis'.

Hölderlin's conception of nature is an organism unfolding to perfection. I have argued that he sees modernity as an important stage of this unfolding, which is characterized by the development of technology through human actions. I have further argued that this means that the 'environmental crisis' of modernity – a side-effect of the development of technology – is also an inevitable stage of this unfolding; it is in the interests of nature. As nature continues to unfold, the disharmony of modernity will be succeeded by a re-conquered harmony. I have argued that Hölderlin's 'saving power' is actual technology, as this seems most consistent with his thought. Heidegger's view, that the 'saving power' is the essencing of technology, seems inconsistent with the positive role of technology in cosmic evolution that is envisioned by Hölderlin.

The reinterpretation I have outlined clearly entails an inversion of the conventional notion of causality in the 'environmental crisis' of modernity. Humanity is conventionally pictured as harming nature. My thesis has shown that for Hölderlin it is nature that is 'harming' humanity. We have been cast aside out of the rapture of nature into a realm of suffering and self-consciousness, with the purpose of developing technology to serve the purposes of the unfolding nature of which we are a part.

We are left with the question of what our attitudes to nature should be, given this reinterpretation of what it means to be a human in cosmic evolution. The answer is simple. As nature is the source of our individual attitudes, our attitudes to nature must be in the interests of nature. Our attitudes, whether they are techno-centric, environmentalist, quietist, or nature-exploitative are all correct for us as individuals, because in the aggregate they fulfil the purpose of nature as a whole.

BIBLIOGRAPHY

Feenberg, Andrew, 'Critical Evaluation of Heidegger and Borgmann', in R.C. Scharff and V. Dusek (eds.), *Philosophy of Technology: The Technological Condition – An Anthology,* Oxford, Blackwell Publishing, 2003.

Gatens, Moira, *Imaginary Bodies: Ethics, Power and Corporeality,* London, Routledge, 1996. Haverkamp, Anselm, *Leaves of Mourning: Hölderlin's Late Work,* New York, SUNY Press, 1996.

Heidegger, Martin, *Existence and Being,* London, Vision Press Ltd., 1956.

Heidegger, Martin, 'The Question Concerning Technology', in R.C. Scharff and V. Dusek (eds.), *Philosophy of Technology: The Technological Condition – An Anthology,* Oxford, Blackwell Publishing, 2003.

Hölderlin, Friedrich, 'Being Judgement Possibility', in J. M. Bernstein (ed.), *Classic and Romantic German Aesthetics,* Cambridge, Cambridge University Press, 2003.

Hölderlin, Friedrich, 'Hyperion', in Eric L. Santner (ed.), *Hyperion and Selected Poems,* New York, Continuum, 1990.

Hölderlin, Friedrich, 'No.41: To his Mother', in Thomas Pfau (ed.), *Friedrich Hölderlin: Essays and Letters on Theory,* New York, SUNY Press, 1988.

Hölderlin, Friedrich, 'The Perspective from which We Have to Look at Antiquity', in Thomas Pfau (ed.), *Friedrich Hölderlin: Essays and Letters on Theory,* New York, SUNY Press, 1988.

Jonas, Hans, *The Phenomenon of Life: Toward a Philosophical Biology,* Illinois, Northwestern University Press, 2001.

Nauen, Franz Gabriel, *Revolution, Idealism and Human Freedom: Schelling, Hölderlin and Hegel and the Crisis of Early German Idealism,* Indiana University Press, 2001.

Peacock, Ronald, *Hölderlin,* London, Methuen & Co. Ltd, 1938.

Pfau, Thomas, *Friedrich Hölderlin: Essays and Letters on Theory,* New York, SUNY Press, 1988.

Scharff, R. C., and Dusek, V., 'Introduction to Heidegger on Technology', in R.C. Scharff and V. Dusek (eds.), *Philosophy of Technology: The Technological Condition – An Anthology,* Oxford, Blackwell Publishing, 2003.

Schmidt, Dennis J., *On Germans and Other Greeks,* Indiana University Press, 2001.

Sloep, Peter B., and Dam-Mieras, Maris C.E. van, 'Science on Environmental Problems', in P. Glasbergen and A. Blowers (eds.) *Environmental Policy in an International Context: Perspectives,* Oxford, Butterworth-Heinmann, 2003.

Spinoza, Benedict de, 'LVIII: To Schuller', trans. A. Wolf (ed.), *The Correspondence of Spinoza,* 2nd ed., London, Frank Cass & Co. Ltd., 1966.

Stone, Alison, 'Irigaray and Hölderlin on the Relation Between Nature and Culture', in *Continental Philosophy Review,* vol. 36, no. 4, 2003. Stone, Alison, *Nature in Continental Philosophy – Week 4, Section V, Friedrich Hölderlin,* [online], http://www.lancaster.ac.uk/depts/philosophy/away mave/408new/wk4.htm
[accessed 25 October 2005].

Taylor, Charles, *Sources of the Self: The Making of the Modern Identity,* Massachusetts, Harvard University Press, 1994.

Unger, Richard, *Friedrich Hölderlin,* Boston, Twayne Publishers, 1984.

Appendix B

How much of man is natural?

The definition of natural is 'present in or produced by nature'. Is it not obvious to anyone who thinks about the question of 'how much of man is natural' that man has been *produced by* nature, and that every fibre of his being and existence is *present in* nature? Surely, a more appropriate question would be: "How could man possibly doubt that he is completely natural?"

In a trivial sense man, as a creator of words, can create a word such as 'natural', and define it in such a way that it excludes man. The word 'natural' can be opposed to *either* 'artificial' *or* 'supernatural'. However, interestingly, the word 'artificial' is defined as 'made by humans; produced rather than natural'. This definition does not refer to man *himself;* rather the word 'artificial' is itself an arbitrary construct, a word of use in human communication because it enables the *productive activities of man* to be referred to. There is no notion in the word 'artificial' that man himself is not natural. Furthermore, there is no implication that in the world itself there is a fundamental division in nature that the word 'artificial' refers to. It is simply of use to man to have a word that labels the results of his productive activities.

The notion of 'human production' is actually a deeply problematic one. If one gives the matter no real thought then the distinction between the 'natural' and the 'artificial' seems to be obvious, but reflection reveals otherwise. It is obviously the case that before the human species evolved nothing was artificial. It also seems obvious that objects such as the Sun and the planet Jupiter are not artificial. But when we focus on the Earth, then it is hard to identify anything that is truly natural. Let us consider a tree that is growing in a rainforest, and a tree that is produced in a human factory. One would be tempted to call the former 'natural' and the latter a 'produced artefact'. However, when one learns that the rainforest tree has the particular attributes that it has because humans soaked the surrounding ground with nutrients and breathed additional carbon dioxide into the vicinity of the tree, then one might have to concede that the tree is an artefact.

Similarly, a human-constructed wigwam-shaped structure composed of branches would be a 'produced artefact', but a single branch lying on the ground under a tree would be considered 'natural'. Even if a human steps on the branch and breaks it the branch would still be considered 'natural'. But there is no difference in kind between modifying a branch by stepping on it, and moving several branches into a wigwam structure. At a larger scale human activities have modified the climate and atmosphere of the entire planet thereby making the concept of the 'natural', as opposed to the 'artificial', largely redundant when it comes to the biosphere of the Earth. So, there is no meaningful distinction *in reality itself* between the 'artificial' and the 'natural'.

The word 'supernatural' is defined as: 'of or relating to existence outside the natural world'. It has to be questionable whether this word has any meaning whatsoever – surely all that exists is the natural world – *nothing* exists outside it. The word 'supernatural' is also used to refer to 'a power that seems to violate or go beyond natural forces', and also 'of or relating to the miraculous'. These descriptions are instructive because they imply that man uses the word 'supernatural' to refer to those parts of the universe around him that he cannot comprehend. So, in the past a total eclipse of the sun would have been referred to as a supernatural event, a miracle. But now such an event is simply considered to be a natural occurrence. This means that the word 'supernatural' delineates man's *understanding* of how the universe works from how it *actually* works; it doesn't imply that in reality parts of the universe are not natural. A complete description of the universe would include solely 'natural' phenomena; there wouldn't be any 'supernatural' phenomena.

So, the word 'artificial' simply refers to the productive activities of man in the world, and the word 'supernatural' can be thought of as delineating the limits of man's understanding of the world. These terms are perfectly compatible with the belief that man is wholly natural, and the belief that all that exists is the natural world. Indeed, if one denies this and asserts that the world, and the Earth, contains natural parts and non-natural parts then one faces a seemingly inextricable problem. One has to try to untangle the complex and intricate way in which the activities of man have modified man's surroundings in order to separate what is natural from what is not natural. If one attempted to do this then one would surely conclude that the entire biosphere of the Earth is *not* natural. It is surely more sensible to conclude that the whole of the Earth is natural, and that *the word 'artificial' refers to things that, whilst produced by man, are still natural*. Alternatively, if one really believes that *the world itself* is split into the natural and the supernatural, then one has to specify exactly what these supernatural entities are. History suggests that there are no such entities because as understanding increases the 'supernatural' gets reclassified as natural.

Nevertheless, the very fact that man can doubt whether he is natural is clearly of interest. To understand the existence of this doubt we need to deconstruct the word 'natural'. Natural has been defined as: 'present in or produced by nature'. The word 'nature' itself has not yet been defined, but is contemporarily defined as: 'the material world and its phenomena'. This definition of nature sheds light on why man could possibly doubt that he is completely natural. It is the contemporary *conception* of nature as the 'material world' that causes man to doubt whether he is 'natural' because man himself doesn't seem to be composed of 'mere matter'.

In the face of this doubt there are two 'naturalistic' options. Firstly, one could hold that the term 'mere matter' is an accurate description of much of the world, but that man is not made of 'mere matter', and therefore that 'mere matter' has the ability to *produce* entities such as man which are not made of 'mere matter'. Secondly, one could hold that the term 'mere matter' is vacuous – there are no parts of the world that fit the term. In other words, one can hold that man's knowledge of 'nature' is still at a primitive level and that the label 'material world' is deeply misleading; it could be the case that all of nature has states analogous to those in man. Both of these options are 'naturalistic' options; both entail that man *is* wholly natural. However, if one's beliefs put one into the first group then it is much more likely that one will consider oneself to *not* be natural.

Most of the above analysis relies on the definition of words – definitions that can help explain why man might not consider himself to be natural, *given* the definition of natural. However, there is a much deeper issue as the definitions themselves are clearly expressions of man's pre-existing conceptual framework. The real issue is why man's conceptual framework in the current epoch is such that he defines nature as the 'material world and its phenomena'. Why do most contemporary humans consider themselves to be opposed to the surrounding world in some fundamental way? Could it be that it is a fundamental characteristic of what it is to be human to conceive of oneself as opposed to the surrounding world?

There are clearly two distinct issues. Firstly, given that the word natural means 'present in or produced by the material world and its phenomena', what could it possibly mean for man to be 'not natural'? Secondly, if man is wholly natural why does he doubt his 'naturalness'? Why does he consider himself to be opposed to the surrounding world? These two questions are obviously closely interrelated because it is the belief in an opposition between man and the surrounding world which leads to the conceptualizing of that world as 'mere matter'. If the belief in an opposition didn't exist then the natural world itself would, no doubt, not be conceptualized as 'mere matter'. Rather, it would be conceptualized in such a way that the attributes of man and world are tightly coupled.

Man's conception of the natural world

It goes without saying that man's conception of the natural world has varied immensely through time, and that it also varies between different cultures in the contemporary epoch. It is the dominant contemporary 'western view' of the natural world which I will focus on here. This view establishes a particularly sharp division between man and world, a division that is so deep that man's naturalness can be seriously doubted. According to the 'western view' the natural world is a clockwork mechanism the activities of which have been increasingly accurately predicted through science. The operations of the non-living world, and much of the living world, are conceived of as thoroughly deterministic and are wholly devoid of qualitative feeling, intentionality and awareness. In contrast, man has 'free will' to act in ways which cannot be predicted by science, and has qualitative feeling, awareness and intentionality.

Of course, there are those who live in the 'west' who don't subscribe to the 'western view'. Some people hold that the entire natural world has intentionality, or qualitative feeling, or awareness, or that quantum physics shows that the entire natural world has freedom. Others argue that man himself doesn't have free will – it is simply an illusion. What is one to make of the claims of these people who oppose the 'western view'? The issues raised are very deep any many seem to be unanswerable. Who could say whether states of aboutness/intentionality exist when atoms interact to form molecules? Who could say whether these interactions themselves entail qualitative feeling? Who could say whether there is any kind of awareness present in these interactions? Who could say whether these interactions actually result from free will? And who could possibly know if every thought that they have ever had is determined, and in theory predictable before they had it?

I take it that no-one has adequate answers to these questions. Therefore, the question needs to be asked as to why the 'western view' always sides with the 'oppositions' – the answers to the questions which lead to an opposition between man and world. This 'siding' clearly says nothing about the natural world *itself* – all it reveals is the way in which man perceives himself in relation to that world.

I obviously do not claim to have answers to the above questions. However, I do not see any good reason why, given that man was produced by nature, that there should be a great chasm between the attributes of man and world. This doesn't mean that man cannot have unique attributes, just as a human eye has attributes that a human finger lacks. It is possible that the high-level of thought that occurs in a human is a unique attribute of man. But, just as the body of man is pervaded by unique attributes, and just as the attributes of mercury are very different to those of helium, uniqueness doesn't entail a fundamental division in reality. It is surely the case that the vast majority of the attributes of man are shared by the entire natural world; whilst every phenomenon in the natural world (including man) also have some kind of uniqueness when analyzed in detail.

It could be argued that the question: "How much of man is natural?" should be replaced with the question: "How many of the attributes of man are not present in the non-human world?" At a first glance this question would seem to provide some kind of an answer to the former question. If it was concluded that man has a plethora of attributes that are not present in the non-human world then this would seem to imply that a large proportion of man is not natural. Whilst, contrarily, if it was concluded that there are hardly any 'unique' human attributes this would indicate that man is very largely natural. However, it is clear that the latter question cannot provide any kind of answer to the former question. Whilst the latter question is a valid question to ask it is comparable to asking the question: "How many of the attributes of mercury are not present in the non-mercury world?" There clearly are attributes of mercury that are not present in the non-mercury world, because it is the presence of these attributes that makes mercury mercury. But it would be nonsensical to conclude from this that the question: "How much of mercury is natural?" is a sensible question to ask. It is simply the case that within the natural world there are differences in the attributes of the various phenomena that exist; some phenomena will closely resemble others, and some will not.

It is time to consider the place of non-human animals. We have seen that 'natural' is defined as 'present in or produced by the material world and its phenomena', and is conceptually opposed to both 'artificial' and 'supernatural'. Given these 'oppositions' what are we to make of the 'naturalness' of other animals. We know that all living things modify their surrounding environment, and that many species of animals are very human-like in their activities. For example, chimpanzees are tool-users, beavers construct dams, and birds construct nests. These activities and modifications of the surrounding world by non-human animals are clearly not 'artificial', because artificial is defined as 'made by humans; produced rather than natural'. They are surely also not 'supernatural', which is defined as: 'of or relating to existence outside the natural world'. These activities are surely wholly natural.

I take it that it would be indefensible to describe non-human animals themselves, or their constructions, as anything other than wholly 'natural'. But I also take it that it would be woefully inadequate to describe beavers and chimpanzees as 'mere matter'; there are close links between the attributes of humans and the attributes of these non-human animals. But, it has been accepted that beavers and chimpanzees, and their activities, are wholly 'natural'. This means that they are wholly produced by or present in the material world; they are either 'mere matter' or the result of the interactions of 'mere matter'. So, as with humans, there is a clear tension here – beavers and chimpanzees are both 'natural' *and* they are 'more than mere matter' at the same time.

This tension gets to the heart of the issue of the relation between man and the 'natural world'. In fact, other animals are an 'intermediary' between man and phenomena such as mercury. It seems easy to assert that mercury is both wholly natural and 'mere matter'. But when it comes to an animal such as a chimpanzee, whilst it is easy to assert that it is wholly natural, it also seems to be correct to assert that it is 'more than mere matter'. This tension gets continued into the realm of man, because man is also surely 'more than mere matter'. It is this tension which leads to the conclusion that maybe man is not wholly 'natural' because the 'natural' is fundamentally 'mere matter'. But if we accept this conclusion then we surely also have to accept that some non-human animals are also not wholly 'natural'. This is surely unacceptable.

What is the alternative? If the similarities between man and certain species of non-human animals are accepted, which they surely should be, then it has to be accepted that if these non-human animals and their activities are wholly natural, then so is man and his activities. Furthermore, we should change our conceptualization and definition of 'natural' by ridding it of the notion of the 'material' world. In other words, we should initially accept our ignorance about the fundamental nature of reality, and then we should conclude that the word 'material' is vacuous. This means that we can then define 'natural' as: 'present in or produced by the world and its phenomena'. This definition quite helpfully rids us of the notion of the 'supernatural'. It also leaves open the possibility that there is a tight coupling between the attributes of man and world. If this conceptualization became the dominant view of the man-nature relationship, rather than the 'western view', then surely man would consider himself to be wholly natural. In the present epoch man doubts his 'naturalness'.

Why does man doubt his 'naturalness'?

It is slightly paradoxical that man can on the one hand talk of the evolution of all species from a common ancestor and the Big Bang, and yet, on the other hand, he can doubt his 'naturalness'. Perhaps this is so because man is that part of nature which *of necessity* considers itself to be not natural. In other words, in the form of man, nature has produced a kind of 'reflective mirror' which enables nature to do things that are impossible without such a mirror. A useful analogy is the hairdresser, who is capable of creating the perfect haircut for her clients without a mirror, but who can only produce a dreadful mess on her own hair without the aid of a mirror. If the hairdresser produces a mirror she will gain the ability to perfectly cut her own hair, just as nature clearly gains abilities through producing man.

If this is right – if man has unique abilities in nature due to not considering himself to be natural – then this means that it is inevitable that man will doubt his naturalness. To be man *is* to consider oneself to not be natural; to be opposed to the surrounding world; to be alienated from the rest of nature. On this view it is inevitable that man will doubt his 'naturalness'.

Concluding remarks

I have argued that man is completely and utterly natural in every fibre of his being. The word 'artificial' is simply a label that is used in human communication to refer to the productive activities of man in the world. The word 'supernatural' simply delineates man's understanding of the world from the way the world actually is.

I have claimed that some species of non-human animals are sufficiently similar to man that if man isn't wholly natural then this implies that these non-human species are also not wholly natural. I have also claimed that it would be futile to attempt to divide the phenomena of the Earth up into the 'natural' and the 'artificial'; the inextricability of the 'artificial'/'natural' division forces one to accept that everything is natural.

I have proposed that a more perplexing question than: "How much of man is natural?" is, "How could man possibly doubt that he is completely natural?" I have suggested that man is that part of nature which inevitably comes to consider itself to not be natural. It is this belief, this doubt, which gives man his unique position in nature. Man and doubt are inextricably linked. Nevertheless, man is wholly and utterly natural.

www.ingramcontent.com/pod-product-compliance
Lightning Source LLC
Chambersburg PA
CBHW060032210326
41520CB00009B/1094